A Primer for
Experimental Evolution

Other Title by the Authors

Methuselah Flies: A Case Study in the Evolution of Aging
ISBN: 978-981-238-741-7

A Primer for
Experimental Evolution

Michael R. Rose
University of California, Irvine, USA

Richard E. Lenski
Michigan State University, USA

Margarida Matos
University of Lisbon, Portugal

Joseph L. Graves Jr.
North Carolina A&T State University, USA

World Scientific

NEW JERSEY · LONDON · SINGAPORE · BEIJING · SHANGHAI · HONG KONG · TAIPEI · CHENNAI · TOKYO

Published by

World Scientific Publishing Co. Pte. Ltd.

5 Toh Tuck Link, Singapore 596224

USA office: 27 Warren Street, Suite 401-402, Hackensack, NJ 07601

UK office: 57 Shelton Street, Covent Garden, London WC2H 9HE

Library of Congress Cataloging-in-Publication Data
Names: Rose, Michael R. (Michael Robertson), 1955–   author | Lenski, Richard author |
    Matos, Margarida author | Graves, Joseph L., 1955–   author
Title: A primer for experimental evolution / Michael R. Rose, University of California, Irvine, USA,
    Richard E. Lenski, Michigan State University, USA, Margarida Matos,
    University of Lisbon, Portugal, Joseph L. Graves Jr., North Carolina A&T State University, USA.
Description: New Jersey : World Scientific, [2026] | Includes bibliographical references and index.
Identifiers: LCCN 2026008321 | ISBN 9789819825400 hardcover |
    ISBN 9789819825974 paperback | ISBN 9789819825417 ebook for institutions |
    ISBN 9789819825424 ebook for individuals
Subjects: LCSH: Experimental evolution
Classification: LCC QH362 .R67 2026
LC record available at https://lccn.loc.gov/2026008321

British Library Cataloguing-in-Publication Data
A catalogue record for this book is available from the British Library.

The cover image is a composite of a sketch from one of Darwin's notebooks and photographs of fruit flies, flasks, and a petri dish with bacterial colonies. The Darwin sketch was rendered from the original by sciencefigures.org under an open design license; the photos of the petri dish and flasks are from the Lenski lab. The overall cover design was produced by Lionel Seow.

For any available supplementary material, please visit
https://www.worldscientific.com/worldscibooks/10.1142/14651#t=suppl

Desk Editor: Brittney Phitojo

Typeset by Stallion Press
Email: enquiries@stallionpress.com

*To our students, colleagues, friends, and family who have supported
us in our efforts to understand how endless forms most beautiful and
most wonderful continue to evolve.*

About the Authors

Michael R. Rose is a Distinguished Professor Emeritus of Biology at the University of California, Irvine. He has been studying the biology of aging since 1976 and has received several scientific prizes, including the inaugural Bacon Prize in 2019. With more than three hundred academic articles and ten academic books, his research on aging is summarized in *Conceptual Breakthroughs in the Evolutionary Biology of Aging*, co-authored with Kenneth Arnold.

Richard E. Lenski is the John Hannah Distinguished Professor at Michigan State University, best known for the Long-Term Evolution Experiment he began in 1988, which continues to this day. The experiment provides a unique record of evolution, offering insights into the dynamics of adaptation, genome evolution, repeatability of evolution, and the origin of new functions. Dr. Lenski is a member of the National Academy of Sciences, American Academy of Arts and Sciences, and American Philosophical Society. He is a past President of the Society for the Study of Evolution, co-founder of the NSF-funded BEACON Center for the Study of Evolution in Action, and recipient of the Friend of Darwin award from the National Center for Science Education for his public-facing work on evolution. He has mentored around 30 graduate students and postdoctoral scientists who now hold faculty positions in universities around the US and the world.

Margarida Matos is Full Professor at the Department of Biology, Faculty of Sciences, University of Lisbon, where she coordinated a Ph.D. program and taught several courses in Evolutionary Biology, including Experimental

Evolution. She is a researcher at the Centre for Ecology, Evolution and Environmental Changes, and for many years led the Local Adaptation in *Drosophila* Research Group. Her research focuses on evolutionary patterns and processes during adaptation to novel environments in *Drosophila* populations, including thermal adaptation and the influence of genetic backgrounds on evolutionary trajectories. She has authored over 50 articles in leading journals and co-edited *Methuselah Flies: A Case Study in the Evolution of Aging.* She is active in manuscript reviewing and serves as editor for *Frontiers in Genetics.*

Joseph L. Graves, Jr. is MacKenzie Scott Endowed Professor of Biology at North Carolina Agricultural and Technical State University, specializing on the evolutionary genomics of adaptation. In 1994, he was elected a Fellow of the American Association for the Advancement of Science (AAAS). His recent honors include the Genius Award (Liberty Science Center), Outstanding Alumnus Award (Public Service, Oberlin College), Patrusky Lecture for Science Writers Conference, and the W.W. Howells Award for best book in biological anthropology, presented by the American Association of Anthropologists (AAA). He is Associate Director of Precision Microbiome Engineering (PreMiEr) NSF Gen-4 ERC and the Director of the NC Amgen Biotech Experience. Dr. Graves is author of *Why Black People Die Sooner, A Voice in the Wilderness, Racism, Not Race* (with Alan Goodman), *Principles and Applications of Antimicrobial Nanomaterials, The Emperor's New Clothes,* and *The Race Myth.* He has also served on the Racial Reconciliation and Justice Commission, COVID Vaccination Task Force, and the Council of Advice on Public Policy (CAPP) of the Episcopal Diocese of North Carolina, and as a science advisor to several theological seminaries through the AAAS Dialogues of Science, Ethics, and Religion (DoSER) program.

Contents

Chapter 5   Genetic and Genomic Responses to Experimental Evolution   57

Chapter 6   Experimental Evolution in Asexual Populations   75

Chapter 8   Conclusions 127

Preface

Around the turn of the millennium, biology received a wondrous gift: whole-genome sequencing. It had been expected that sequencing the entire genomes of nematodes, fruit flies, and humans would rapidly reveal the genetic foundations of both function and disease. Indeed, that had been the assumption that motivated the substantial funding of this new field of "genomics" in the United States, the United Kingdom, China, and Japan.

This confidence grew out of the prevailing cell and molecular reductionism that held sway across many university biology departments and especially biomedical institutes. In such reductionism, the central goal was to identify specific genetic changes that, in turn, led to specific phenotypic changes. The implicit preference was for single genes that produced dramatic phenotypic transformations, as in the so-called human genetic diseases and large-effect Mendelian mutants. But late 20th century genetics was becoming more amenable to the possibility that several genes might jointly contribute to specifying important phenotypes, as they do even for eye color in both humans and fruit flies (Graves, 2023).

Once whole-genome sequences were available, it became clear that the functional significance of genetic variation across the entire genome was not so easily determined. Genome-wide association studies ("GWAS") were the most powerful scientific technology used to decipher genomes in the first two decades of the 21st century. But even when GWAS could identify specific genetic variants that affected specific phenotypes, those variants typically accounted for only a small fraction of the inheritance of such phenotypes. Moreover, those variants often spanned multiple genes with many differences in sequence associated with each variant, making it impossible to disentangle their effects. One of the best-studied biological traits, human height, was shown to be affected by many sites in the human genome (Conery & Grant, 2023). Yet all together, early studies of those sites accounted for only a minority of the heritable variation in height (e.g., Yang et al., 2011). Fortunately, more recent studies using much larger sample sizes (>1,000,000 individuals measured for height) have shown that >12,000 single-nucleotide polymorphisms ("SNPs") across 7,209 genetic loci (~21% of the human genome) can explain around 90% of the SNP-based heritability of human height (Yengo et al., 2022). Thus, GWAS has been found to be a rather inefficient tool, because it requires very large sample sizes to reveal the genome-wide foundation of most traits of biological or medical interest.

This outcome has led to a struggle to develop methods that could more efficiently parse the functional architecture of life, from genes and transcripts to metabolites and whole-organism functions. The conventional methods of 20th-century biology, founded chiefly on the study of standing genetic variation and new mutations that typically disrupt biological functions, are no longer adequate for the genomic biology of the 21st century.

In this primer, we offer experimental evolution as a contender for a leading role in the biology of our time. For those unfamiliar with the term "experimental evolution," it does not refer to any and every type of data collection in evolutionary biology. Much of the data collected by evolutionary biologists concerns the "natural history" of life's unfolding, whether at the level of fossils or DNA sequences. That type of empirical research has revealed much of great interest, and it will continue to illuminate the "what happened" questions of biology.

Experimental evolution is instead a field that uses strong-inference methods to probe the basic mechanisms that underlie evolution, function, and even disease (e.g., Basu et al., 2024). It conducts experiments in which evolution is forced to yield its secrets, and it has the power to probe deeply into the genomic machinery on which life depends and by which, in the words of Charles Darwin, "endless forms most beautiful and most wonderful have been, and are being, evolved."

Experimental evolution was an emerging field without an accepted name for most of its development in the 20th century. Early experiments in the field suffered from small scale and unreliable controls. Some of those experiments were conducted in natural populations that were subject to confounding fluctuations in migration and climate. But in the 1980s and 1990s, the field began to develop a tradition of large-scale and rigorous experiments. Multiple evolving populations were employed, and appropriate control populations were incorporated into the experiments. From dozens of generations prior to 1980, experimental evolution projects expanded in their duration from hundreds of generations to thousands and even tens of thousands of generations.

With the advent of whole-genome sequencing in the late 1990s and early 2000s, and with the rapidly declining costs of acquiring those data, experimental evolution studies turned from the characterization of evolution in a few candidate genes to sequencing the entire genomes of large samples taken from multiple replicate populations. By 2020, it was apparent that the procedures of experimental evolution could be combined with genome-wide sequencing to yield scientific findings of remarkable power. After two decades of frustration with the limited power of methods like GWAS, the 21st century now has a scientific tool that can address many unsolved problems in biology at the scale of the entire genome.

Fortunately for biologists in training, the methods and motives of research in experimental evolution are neither tricky nor opaque. In this respect, experimental evolution is like the experimental physics that started in the hands of Galileo and Newton in the 17th century. Experimental physics added carefully controlled and reproducible procedures to address key questions in physics at that time and ever since. It supplemented the astronomy and mechanical engineering that had previously been the proving grounds for the theories of physicists. But more importantly, experimental physics was used to perform strong-inference tests of alternative theoretical models for phenomena like motion and the scattering of light.

As experimental evolutionists, we see ourselves as continuing the traditions of experimental physics. We do not focus on historically unique events, like the evolution of mammals or the properties of a single mutation of large effect. Instead, we perform experiments that can be, and often are, repeated across laboratories and time. We are interested in reproducible patterns obtained under well-defined conditions. Further, we strive to achieve whatever levels of replication, control, and scale are required to produce strong inferences about the biological phenomena we study, just as physicists do when researching physical phenomena.

Experimental evolution is not a field that emphasizes narrative appeal or seeks to inspire feelings of awe. Evolutionary biology has had many superb storytellers about the history of life, including its innovations and extinctions. These stories inform and delight the curious. By contrast, experimental evolution seeks to cut through the tangled knots of biology's biggest unsolved puzzles. Experimental evolution does not satisfy itself with ideas or theories that are intuitively plausible or easy to explain. Instead, as we will show, it seeks to confront some of the most fundamental and challenging biological questions.

Acknowledgments

We thank Christopher Davis and Brittney Phitojo for their editorial assistance and skillfully guiding this book to completion. We thank Misty Thomas for preparing many of the figures in this volume, along with Michael Wiser, Olivier Tenaillon, and Benjamin Good, who each prepared a single figure, as well as Kenneth Arnold who prepared Figures 7.3 and 7.4.

We extend our heartfelt thanks to the countless individuals — technicians and lab managers, undergraduate and graduate students, postdoctoral researchers, and collaborators from around the world — who have worked with us through the years and decades of our evolution experiments.

We greatly appreciate the support that we have received from various funding sources. Michael Rose is grateful for 50 years of support provided collectively and sequentially by the British Commonwealth Scholarship program administered by the British Council, the NATO Postdoctoral Science Fellowship Program, the Canadian University Research Fellowship Program administered by Canada's Natural Sciences and Engineering Research Council (NSERC), and research grants from NSERC, Dalhousie University, the U.S. National Institutes of Health, the U.S. National Science Foundation (NSF), and the University of California. Richard Lenski wishes to acknowledge, in particular, the NSF for long-term support of the LTEE, a USDA Hatch Grant that pays a portion of his salary, and the John A. Hannah professorial endowment at Michigan State University. Margarida Matos acknowledges funding from the Portuguese Foundation for Science and Technology (FCT) as well as the Centre for Ecology, Evolution and Environmental Changes (CE3C) and the Faculty of Sciences of the University of Lisbon for financial and logistic support through the years. Joseph L. Graves Jr. acknowledges funding from his MacKenzie Scott Endowed professorship that pays a portion of his salary, as well as past NSF support for his experimental evolution studies of metal resistance in microbes.

On a personal level, we thank our families for putting up with us during the many years of long hours in the lab, writing papers, attending conferences, and more. Michael Rose is grateful for his assorted wives and children, especially Darius, Alexander, and James. Richard Lenski gives special thanks to his late parents, Gerhard and Jean Lenski; to his wife, Madeleine; and to his three children, Daniel, Shoshannah, and Natalie. Margarida Matos is grateful to her husband, Pedro and two children, Alice and Francisco, for all their support and for understanding the many hours that were stolen from shared company. Joseph L. Graves Jr. thanks his late parents, Joseph and Helen, who made it possible for him to assume the mantle of "Black Darwin," as well as his wife, Sue, and sons, Joey and Xavier, for their ongoing support of his work.

Chapter 1

Big Questions About Evolution

Charles Darwin's (1859) theory of evolution by natural selection is considered one of the most important scientific achievements of all time. Darwin gave us the broad outlines of how evolution occurs in sexual organisms, but we now know that his original formulation was flawed in several respects. The first of these flaws was the notion that inheritance involves blending (Mayr, 1982). With blending inheritance, offspring combine the attributes of their parents by some type of mixing of ill-defined fluids or particles, much like mixing paints of different colors. Moreover, these fluids or particles, which Darwin called "gemmules," might even be shaped by an organism's experiences, leading to the inheritance of parentally acquired characteristics (Mayr, 1982). Blending inheritance was widely assumed by the biologists of Darwin's time, and his invocation of it was not controversial in the 19th century. This error was corrected after the rediscovery of Mendel's research around 1900, which in turn led to the synthesis of genetics with natural selection during the first few decades of the 20th century (Bowler, 1989).

Darwin's second error was his assumption that adaptive evolution invariably proceeded slowly. He knew that evolution could produce "sports," physically deformed offspring that did not resemble their parents. But, like Aristotle before him, he assumed that such monstrosities would be eliminated by the action of natural selection. For natural selection to produce adaptation, Darwin believed that it must proceed through the slow accretion of small benefits molded by the action of natural selection. Indeed, Darwin wrote in *The Origin* (1859): "We see nothing of these slow changes in progress, until the hand of time has marked the long

1

lapse of ages, and then so imperfect is our view into long past geological ages that we only see that the forms of life are now different from what they formerly were."

Darwin's third major error was that he never considered how evolution might proceed in asexual organisms, chiefly microbes (Kofler & Maloy, 2012). During the first half of the 19th century, microbiology was a controversial field. The status of viruses, bacteria, and protozoa as life forms remained in question, in part because microscopy was inadequate to the task of determining the nature of microorganisms and their means of reproduction. In any case, the burgeoning of experimental microbiology and improved microscopy in the late 19th and early 20th centuries firmly established the existence of microbial life. However, the nature of genetic inheritance in microbes remained controversial until the middle of the 20th century (Lenski, 2017b). As most microbes do not reproduce sexually in the same manner as plants and animals, it also became clear that asexual reproduction was a common feature of life.

It took some time to revise the theory of evolution with respect to the second error, the assumption of gradual adaptation. The rapid evolution of darker pigmentation among European moths from 1840 to 1950 was one of the first cases in which evolution by natural selection was shown to be capable of great speed (Ford, 1964). In a medical context, the rapid evolution of bacterial resistance to antibiotics became increasingly clear in the second half of the 20th century (Levy, 2002). In time, laboratory experiments with bacteria, unicellular eukaryotes, fungi, animals, and plants all demonstrated the possibility of rapid adaptation (Garland & Rose, 2009).

However, before these cases of rapid evolution were understood, Darwin's idea of gradualism had become a central tenet in what was known as the "modern" or "Neo-Darwinian" synthesis. Julian Huxley's *Evolution: The Modern Synthesis* (1942) proposed that gradual evolution could be explained in terms of many minor alleles brought together by sexual recombination. Natural selection was supposed to act upon and accumulate these imperceptibly small variations over long periods, eventually culminating in observable differences between separate lineages and the process of speciation. Since Huxley's time, biologists have gained a fuller understanding of evolutionary mechanisms, including how mutations produce genetic variation, the interaction between random drift and natural selection, the roles of migration and nonrandom mating, and so on.

In any case, it eventually became apparent that experimental approaches to evolutionary questions might advance the field more quickly than most biologists had imagined prior to 1970 (see Wright, 1977). Among the big questions waiting to be explored and answered were those concerning (i) the evolution of sexual reproduction; (ii) the evolution of aging; (iii) evolvability and the evolution of genetic processes; (iv) the tempo of evolutionary change; (v) the origin of life; (vi) the evolution of multicellularity; and (vii) the processes of speciation and adaptive radiation.

1.1. Some of the Big Questions

1.1.1. *Evolution of sex*

Questions and hypotheses about the origin of sex and the maintenance of sexual reproduction have a long history (e.g., Geddes & Thompson, 1908). One of the first cogent explanations for sexual reproduction came from H. J. Muller (Muller, 1932). He pointed out that sexual recombination would bring together favorable variants at different chromosomal locations much faster than could occur in organisms without sex. He reasoned that, in asexual organisms, two beneficial mutations at different sites in the genome would have to spread sequentially because the rate at which favorable mutations occur is very low, making the simultaneous occurrence of both mutations in a single individual highly unlikely. By contrast, sexual recombination can bring together two beneficial mutations more quickly, even when they first appear in different individuals.

A second potential advantage to sexual reproduction occurs when populations are small enough to be affected by genetic drift. Muller (1964) reasoned that, over evolutionary time, a small asexual population would lose those lineages with the fewest deleterious mutations by random drift, a process now known as Muller's Ratchet (Maynard Smith, 1978). As a result, small asexual populations would accumulate more and more deleterious mutations, eventually leading to their extinction (Bell, 1988). By contrast, sexual reproduction would continually reconstitute individuals with few deleterious mutations via outcrossing and recombination.

Third, sex may be maintained because recombination in populations with standing genetic variation produces greater functional diversity among offspring in each generation than can be produced by clonal reproduction. This diversity is particularly useful for species in environments that vary in time and space, as they usually do (Maynard Smith, 1978). One environmental factor that is thought to particularly favor sex is parasitism (Hamilton, 1980). In an asexual population with little genetic diversity, a parasite might drive the population to extinction before it can evolve resistance. Sexual populations might have enough variation to produce at least some offspring that are sufficiently resistant to parasites, so they can survive to reproduce (Seger & Hamilton, 1988).

In any case, for sex to evolve and be maintained by natural selection, the benefits of sexual recombination must outweigh its costs (Maynard Smith, 1978). All else being equal, and assuming that males merely provide gametes (and play no role in, e.g., offspring care), then asexual females should produce twice as many daughters as sexual females. There are often other costs to sexual reproduction as well, such as the need for structures and behaviors, the risks of exposure to predation while seeking sex, and so on. All of the costs listed above have led to the paradoxical proposal that sex might actually not be functional but rather the result of ancient and long-sustained evolutionary parasitism, in essence by males or the genetic elements that led to males (Hickey & Rose, 1988; Krieber & Rose, 1986; Rose, 1983).

Thus, the evolution of sex is a big, unanswered question in biology. Fundamental questions about its benefits, its origins, and its maintenance remain unsettled. We believe this question illustrates the diversity of hypotheses that might be tested using experimental evolution and that could provide stronger inferences than those supplied by natural history and mathematical theory.

1.1.2. *Evolution of aging*

The progressive deterioration of individual animals, including humans, as they age is a classic problem of biology, one dating back to the origins of civilization. It occurs even when individuals are raised under healthful conditions. Furthermore, aging represents a failure of age-specific functionality that starts in middle age and then tends to worsen. Aging is a puzzle for evolutionary biology because one might naively expect that natural selection would have eliminated such failures in functionality.

Evolutionary theorists proposed several scenarios for the evolution of aging in the middle decades of the 20th century (e.g., Medawar, 1952). These ideas were initially expressed in words, then formulated mathematically and shown to be capable of producing the evolution of aging, providing there is genetic variation with age-specific effects on those functions that govern performance and survival (e.g., Charlesworth, 1980). Unfortunately, merely observing the effects of new mutations and the patterns of standing quantitative genetic variation proved to be relatively poor tests for evaluating the validity of alternative evolutionary theories of aging (Rauser et al., 2009).

1.1.3. *Evolvability*

Evolvability is the capacity of biological systems to generate heritable phenotypic variation, which is the requisite "fuel" that powers adaptation by natural selection. There are many fundamental questions related to evolvability. Does adaptive evolution proceed by gradual or punctuated change? Are there limits to adaptation? How do rates of mutation and recombination impact the rate of adaptive evolution? Can the rates of mutation and recombination themselves evolve? What role does phenotypic plasticity play in adaptation? How do organisms respond to multiple selection pressures? If populations evolve in similar or identical environments, will they adapt in the same way, or might they instead follow different paths and evolve different solutions (Gould, 1989; Losos, 2018)?

1.1.4. *Gradual versus punctuated evolution*

Based on the fossil record, Gould and Eldredge (1993) argued that evolution could exhibit long periods of stasis, with no or only gradual change, that were punctuated by periods of rapid change. They offered this theory of "punctuated equilibrium" as an alternative to what they called "phyletic gradualism." Some critics have suggested that this was a false dichotomy and that the theory of punctuated equilibrium conflates several questions, such as whether evolution is gradual or pulsed, whether trait evolution occurs primarily during speciation (as lineages diverge) or within a single lineage, whether changes during speciation are adaptive or neutral, and whether higher-level "species selection" is important in shaping patterns of diversity (Pennell et al., 2014). Of course, one of the difficulties of testing theories about the pace of evolution that rely on the fossil record is the incompleteness of that record. While evolution experiments cannot be performed at the spatial and temporal scales reflected in the fossil record, they can address general questions about the uniformity of rates of evolution and the factors that can cause variation in those rates.

1.1.5. *Origin of life*

Some of the greatest evolutionary questions concern the origins of life and multicellularity and the processes of speciation and adaptive radiation. All but the staunchest creationists accept the fact that evolution can produce changes within species. Indeed, evolution in action has been directly observed in cases like the evolution of pesticide and antibiotic resistance as well as in the emergence and spread of pathogens such as HIV and SARS-CoV-2. On the other hand, the origin of life and the evolution of new species have been harder to observe, and therefore they receive more skepticism.

There are useful models and illustrative experiments that suggest how chemical processes could have led to the emergence of living systems defined by the criteria of organization, metabolism, genetic codes, and self-replication (Ellington & Szostak, 1990; Higgs, 2017; Miller, 1953). A key insight is that natural selection, in the sense of differential replicative success of molecules, is likely to have played a critical role in generating pre- and proto-biotic forms (i.e., incipient life), rather than beginning only after the origin of what we would now recognize as life.

1.1.6. *Evolution of multicellularity*

The emergence of multicellularity faces a problem similar to one of the hurdles for the evolution of sex. A cell that replicates and then divides into two daughter cells experiences an increase in its abundance of 100% each generation. If the daughter cells instead fail to separate after division, then they remain a single organism that, in a sense, failed to reproduce. Nonetheless, natural selection could favor such a system if the benefits of the daughter cells staying connected outweigh the lost reproductive opportunity. Such benefits might involve specialization of cells, with only some specializing as the "germ line," giving rise to the next generation, while other cells perform different functions, such as garnering more resources. Being larger might also avoid death from predators or allow an increased number of offspring that overcomes the delayed reproduction (Herron et al., 2022).

Various forms of multicellularity have evolved repeatedly, including in some bacterial lineages and other organisms that are usually unicellular. In some cases, multicellularity occurs by the aggregation of previously separate cells; in other cases, cells remain connected after cell division. Multicellularity has evolved independently many times even in eukaryotes (Grosberg & Strathmann, 2007), although all multicellular animals (metazoans) trace back to a singular origin at least 600 million years ago (King, 2004).

What we usually regard as complex organisms are those that have multiple tissue types and organ systems that perform different functions, thereby facilitating survival and reproduction in variable environments. While biochemical complexity arose in unicellular systems, the specialization of tissues and organs began with the evolution of multicellularity. Even so, unicellular organisms dominate in terms of their total numbers, although multicellular terrestrial plants dominate in terms of biomass. As Stephen Jay Gould (1996) has said, "bacteria are—and always have been—the dominant forms of life on earth."

1.1.7. *Speciation and adaptive radiations*

Darwin (1859) called the process by which new species arise "the mystery of mysteries," and data bearing on the formation of new species by natural selection was once called "Darwin's missing evidence" (Kettlewell, 1959). Neo-Darwinian theory proposed that speciation in sexual species is driven by local adaptation combined with a cessation of gene flow, particularly when geographical separation occurs (Mayr, 1963). In principle, these conditions could be studied in a laboratory setting. Early experiments with *Drosophila* were suggestive, and they provided some evidence for selective mating between males and females adapted to different conditions. However, the experiments were not sustained for enough generations to produce the level of reproductive isolation required to claim speciation (e.g., Dodd, 1989; Millicent & Thoday, 1960; Rundle et al., 2005).

The biological species definition does not hold for asexually reproducing organisms like bacteria (Cohan, 1994), for two reasons. First, mating is unnecessary for asexual organisms, rendering the notion of reproductive isolation meaningless. Second, many bacteria can nonetheless share genes (called "horizontal gene transfer") via parasexual processes like conjugation (Shapiro et al., 2012), and these exchanges sometimes occur even across widely divergent species. As a consequence, strains belonging to the same species can have core genomes with high DNA sequence similarity, yet they can also differ by hundreds of genes that are present in one strain but absent in another. Conversely, some species and even genera have been described based on certain phenotypes (e.g., pathogenicity) that resulted from the horizontal acquisition of a relatively small number of genes (Chaudhuri & Henderson, 2012). While species of bacteria were historically defined by their morphology, biochemistry, physiology, and ecology, genomic data now play an important role, although the resulting boundaries and cutoffs may be rather arbitrary. Similar issues occur for viruses, especially when they invade and adapt to new hosts. The SARS-CoV-2 virus, which caused the COVID-19 pandemic, provides a dramatic example of this process (Markov et al., 2023).

Closely aligned with speciation is the concept of adaptive radiation, whereby a single ancestral lineage evolves over time into multiple species that use their environment in different ways. Adaptive radiations are often associated with the colonization of islands and subsequent divergence as different lineages exploit distinct resources and fill different niches, as in the case of the adaptive radiation of Darwin's finches in the Galapagos Islands (Grant & Grant, 2014). Another example is the diverse assemblage of cichlid fishes in Lake Victoria and other African lakes, where hundreds of ecologically and morphologically distinct species have evolved within a few million years (e.g., Brawand et al., 2014). Anole lizards that inhabit the Greater Antilles in the Caribbean Sea provide yet another example, one that is particularly interesting because the lizards underwent parallel radiations into ecologically and morphologically distinct species on several islands (Losos, 2009).

As we will see in later chapters, experimental evolution studies have allowed researchers to observe such phenomena as a bacterial strain diverging into multiple lineages that can coexist because they use the available resources in different ways, viruses evolving to use novel hosts, fruit flies evolving to live longer, and much more.

Chapter 2

The What and Why of Experimental Evolution

Modern biologists are divided between those who mostly study cells and molecules in labs and those who do field studies or experiments with whole living organisms. Beginning in the second half of the 20th century, research universities began to place more emphasis on the former than the latter (Yoxen, 1982). However, most modern evolutionary biologists have also adopted the tools of molecular biology while studying living populations of insects, bacteria, and other organisms. Experimental evolutionary biologists belong to this latter group.

2.1. Other Methods of Evolutionary Research

Many evolutionary biologists collect blood or other tissue samples from wild populations, then sequence their genes or transcripts, hoping to interpret patterns of variation within and among populations and species. Other evolutionary biologists take entire organisms from the wild, then bring them into the lab to study their anatomical and physiological properties. This approach is broadly called the *comparative method* because it focuses on comparing the properties of organisms belonging to different populations or species. Such comparisons were once, and perhaps still are, the bulk of empirical research in evolutionary biology. In fact, this approach was the main (though not only) method that Darwin used to infer evolution, and such comparisons were employed profusely throughout *The Origin of Species* (1859). Of course, the comparative method is the primary approach when the goal is to trace phylogenetic histories. However, our intent here is to discuss the various approaches that are used to understand evolutionary processes, not the patterns that have resulted from the action of those processes in the past. In this context, the comparative method must infer evolutionary processes from patterns of differentiation, and it therefore must implicitly assume that the ancestral states of populations and species can be deduced from extant organisms or fossil forms.

Other evolutionary biologists study patterns of heritable variation among individuals from living populations in the wild, on farms, in patient populations, or in laboratories. This field grew out of biometrics, an approach to quantifying heritable variation that was started in the late 19th century by Francis Galton, Karl Pearson, and others. In the 20th century, it became known as *quantitative genetics* (Falconer & Mackay, 1996). Quantitative genetics has been foundational for animal and plant breeding, contributing to the abundance of food that sustains many contemporary human populations. With the development of molecular tools during the 20th century, evolutionary biology turned to other measures of genetic variation, from single enzymes to broader polygenic analyses. Adding these tools allowed evolutionary genetics to expand its questions and approaches (Avise, 2014). The advent of whole-genome sequencing has led to increasingly sophisticated methods, including genome-wide association studies (GWAS) and genome–environment associations (GEA; Lasky et al., 2023).

A common approach used by many quantitative and experimental population geneticists has been to take organisms from wild populations, bring them into the lab, and rapidly inbreed them. (Such inbreeding may proceed by self-fertilization in plants or by mating siblings in animals, generation after generation.) This inbreeding is done to prevent selection in the lab from eliminating the range of genetic variation present in the wild populations from which the organisms were sampled. Inbreeding can also simplify certain genetic analyses. In any case, these inbred lines are sometimes claimed to be the best representation of naturally occurring genetic variation, short of massive direct sampling of the DNA of wild populations.

Computer simulations and theoretical modeling also play an important role in generating and refining evolutionary hypotheses that can be tested empirically. These mathematical tools were critical for the development of the Neo-Darwinian synthesis, which provided the theoretical foundations to understand the interactions of evolutionary processes, including mutation, selection, genetic drift, and migration (Fisher, 1930; Wright, 1931; Wright, 1932).

Experimental evolution is a distinct and increasingly used empirical approach in evolutionary biology. It can be broadly described as conducting real-time evolution experiments under defined and reproducible conditions. Such experiments allow natural selection and other evolutionary processes to occur under controlled conditions with appropriate levels of replication to achieve strong scientific inferences.

In what follows below, we will better delineate the "What" of experimental evolution in general terms. We will then turn to the "Why" of experimental evolution, motivating our discussion using the study of the evolution of aging, one of the big evolutionary questions that has been studied using all the various approaches of biology.

2.2. What is Experimental Evolution?

There are four essential features for conducting experimental evolution: replication, controls, genetic variation, and observing change across many generations. We will describe and discuss each requirement separately.

2.2.1. *Replication*

In general terms, an evolution experiment requires multiple populations that experience the same selective regime and, in some experiments, the same demographic regime (e.g., population size). In the early days of experimental evolution (prior to the 1980s), it was common for a single population to be subjected to a particular selection regime for a number of generations. The central problem with this lack of replication is that it conflates unique accidents of mutation and genetic drift with the action of natural selection.

The basic unit of evolutionary biology, both theoretically and experimentally, is the population. Evolutionary theory makes useful predictions about the properties of populations, but not necessarily about their constituent individuals. The initial failure to include replication at the level of populations meant that inferences based on some early evolution experiments relied on single instances. Ernest Rutherford was fond of denigrating biology with the epithet that a scientific field is either physics or stamp collecting. But studying a single population's evolution does not even rise to the level of stamp collecting. It is more akin to the study of a single stamp, such as the "Inverted Jenny." Thus, the inclusion of replicate populations, each independently experiencing the same selective regime, is fundamental to experimental evolution. The same requirement holds when analyzing variation in outcomes, such as changes in the reproductive rate or the number of mutations that have accumulated. In some experiments, the resulting variance in some property among the replicate populations can be as important as changes in mean properties (Lenski, 2017a).

2.2.2. *Controls*

Controls in experimental evolution are populations that are *not* undergoing the selection or demographic regime of interest. There are two main kinds of useful controls.

The best controls are reconstituted populations that are identical to the populations from which the experimentally evolved populations were derived. In the case of microorganisms, like bacteria, samples of the founding ancestral strain can be frozen. Phenotypic and genetic properties of the experimentally propagated populations can then be compared to thawed samples of their frozen ancestors. Some other organisms also allow freezing, such as nematodes that are used in some evolution experiments. Dormant seeds can serve a similar purpose for evolution experiments with plants. In experimental populations that are founded by hybridizing fully inbred laboratory lines (a common practice in *Drosophila* experimental evolution), a later hybridization procedure can be conducted with the same inbred lines to reconstitute an approximation of the ancestral state of the experimental populations.

Many organisms are not amenable to being frozen and revived or stored as seeds. In some cases, the most appropriate controls are outbred populations that have achieved approximate evolutionary stasis in the culture regime that was imposed on them prior to the derivation of the experimental populations. Ideally, these controls faithfully represent the ancestral state of the populations before the new regime. In any case, when using such organisms, one should maintain parallel control and experimental populations and perform simultaneous assays to analyze the evolutionary changes in the traits of interest over time.

2.2.3. *Genetic variation*

It is foundational to evolutionary biology that evolutionary change requires genetic variation. Without genetic variation, an inbred or genetically homogeneous line may show some fluctuations in its phenotypes, especially if the environment changes. Indeed, such strictly phenotypic changes may be instigated precisely because an investigator imposes different environmental conditions on the control and treatment populations. However, there can be no Darwinian evolution without genetic variation. There are three distinct kinds of genetic variation that can fuel Darwinian evolution: mutation, parasexual recombination, and Mendelian recombination, or "sex."

Mutation is the ultimate source of genetic variation. Mutations can arise by single base-pair changes during DNA replication. They can also arise by insertions or deletions of one or more base pairs as well as from inversions or other chromosomal rearrangements. In large populations, such as microbes that can number in the millions or even billions in a small volume, new mutations can be the sole and sufficient source of genetic variation for experimental evolution.

Parasexual processes like plasmid-mediated conjugation and virus-mediated transduction can introduce more extensive genetic variation than *de novo* mutations. Such processes are common among many bacterial species in nature, and they also occur in some microbial eukaryotes like yeast. A complicating feature of parasexual processes is that the vectors of genetic change, namely plasmids and viruses, can themselves evolve. In effect, there is the potential for a coevolutionary process that involves both the microbial host genome and the vector genome.

Mendelian recombination, in the context of eukaryotic sex, shuffles existing mutations distributed across entire genomes. Eukaryotic genomes are typically split into multiple chromosomes that segregate independently during meiosis. These chromosomes, in turn, can undergo crossover events that further shuffle the genetic variants, or alleles, within chromosomes. Sexual populations thereby generate additional heritable variation in traits whenever there are alleles with functional effects to be shuffled, like cards in the games of poker and bridge, to produce innumerable combinations.

2.2.4. *Observing change across many generations*

Perhaps the most distinctive feature of experimental evolution is that data are collected across many generations, thereby allowing one to quantify the dynamics of evolutionary processes as well as the patterns generated by those processes. As the experimental populations evolve through the generations, one can observe the cumulative effects of natural selection, mutation, recombination, and genetic drift. These effects can lead to increasing differentiation of the populations over time, relative to both their ancestors and one another. Thus, the power of experimental evolution tends to increase with the number of elapsed generations. The resulting differentiation of replicated populations makes statistical inference far more robust. Unlike the variances and covariances among individuals studied by quantitative genetics, the power of experimental evolution depends on the difference of means across populations (ancestor versus evolved, or control versus treatment) and the associated variance among populations. Changes in the phenotypes generated by experimental evolution can also be valuable for understanding biological functions (e.g., Swallow et al., 2009).

2.3. The Why of Experimental Evolution

We will now illustrate the motivating advantages of experimental evolution using one of the Big Questions, namely that of aging. A century ago, aging was one of the great unsolved problems of biology: Why do organisms eventually experience diminished performance and increased risk of dying as they age?

2.3.1. *Evolutionary theory of aging*

The evolutionary theory of aging was first developed verbally (e.g., Medawar, 1952) and then mathematically (e.g., Charlesworth, 1980). It is now considered one of the most successful applications of evolutionary theory over the last 50 years, at least among evolutionary biologists.

Briefly summarized in words, the evolutionary theory of aging proposes that aging is the result of the declining force of natural selection during reproductive adulthood (see Figure 2.1), as first shown

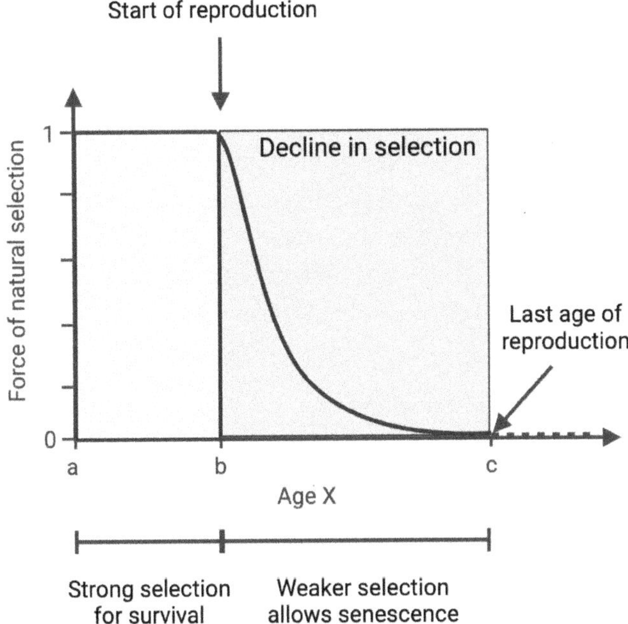

Figure 2.1:   The strength of natural selection declines after organisms reach reproductive maturity, leading to the evolution of senescence. This was first shown mathematically by Hamilton (1966). Aging will be absent when there is no reproductive adulthood, as in the case of species that reproduce solely by symmetrically fissile reproduction. Thus, organisms that reproduce by growing and then dividing into two equal progeny do not age; all that reproduce by asymmetrical means will experience aging.

mathematically by Hamilton (1966). Aging is thus expected to be absent when there is no reproductive adulthood, as in the case of species that reproduce solely by symmetrically fissile reproduction (i.e., when an organism reproduces by growing and then dividing into two equal progeny). Otherwise, aging is expected to be universal.

There are two specific evolutionary genetic mechanisms that arise from the declining force of natural selection. The first is *mutation accumulation*, which results from alleles that have deleterious effects only sufficiently late in life while having no early-acting effects whatsoever. Such allelic variation would accumulate over evolutionary time because it is effectively neutral with respect to Darwinian fitness, which depends on survival and reproduction at earlier ages. The second mechanism for the evolution of aging is *antagonistic pleiotropy,* which results from alleles that have benefits early in life coupled with deleterious effects that are manifest only much later in life. Such alleles are expected to be favored by natural selection, again because Darwinian fitness is primarily determined by the early life effects. These two evolutionary mechanisms are not mutually exclusive, as they can act simultaneously at different genetic loci.

Our chief interest here is to show how evolutionary biology has tested this theory empirically. There are important contrasts in the power of the several empirical approaches used to test its theories.

2.3.2. *Comparative biology of aging*

There are two main contexts for the application of comparative biology to the evolution of aging. The first, and most successful, has been testing the theoretical prediction concerning the presence versus absence of aging, depending on the reproductive mode that species use. The evolutionary theory of aging predicts that strictly and symmetrically fissile species should not exhibit aging. This prediction has been tested multiple times (e.g., Bell, 1984; Martínez, 1998), and it has always been met, to the best of our knowledge. [A complication arises, however, in at least some species that undergo fissile reproduction, because the physical material that constitutes cells may be damaged and distributed unequally to the daughter cells, creating an asymmetry that can lead to an aging-like degenerative process (Rang et al., 2011).] The evolutionary theory of aging also predicts that species that never undergo symmetrically fissile reproduction should invariably age. To the best of our knowledge, this prediction has also been met in well-studied species living in captivity for at least part of their reproductive adulthood (Comfort, 1979; but see Mueller et al., 2011). So, in this respect, comparative studies have been a valuable proving ground for the evolutionary theory of aging.

But problems arise when applying the comparative method to test antagonistic pleiotropy as an evolutionary genetic mechanism for aging. It is generally the case that mammalian species, for example, can be arranged along an axis that features small, early reproducing, and fast-aging species at one end with large, later-reproducing, and slow-aging species at the other (e.g., Stearns, 1976). The proverbial contrast is mice versus elephants. At first glance, this arrangement seems to imply a trade-off between early reproduction and later function, in conformity with antagonistic pleiotropy. Yet there are species that clearly do not fit this comparative pattern. Many small species of bat, for example, can live a long time. Flying birds can produce many more offspring than elephants, yet some bird species can live almost as long as elephants. Thus, comparative patterns of aging have proven to have limited value in evaluating the genetic mechanisms for the evolution of aging. Macroevolutionary patterns observed across distantly related taxa have little power and can even be misleading with respect to what happens at the level of the populations, where evolution really takes place.

2.3.3. *Quantitative genetics of aging*

There are two types of quantitative genetic data that have been brought to bear on the evolutionary genetics of aging: age-dependent patterns of genetic variance among life-history traits and patterns of genetic correlation between early and late life-history traits.

Brian Charlesworth's (1980) theoretical analysis of the process of mutation accumulation was originally thought to imply an increase of genetic variances at later ages for life-history traits, like age-specific survival and fertility. But such increases have not been generally found (e.g., Rose & Charlesworth, 1981a), though they have been seen in a few cases (e.g., Kosuda, 1985). Furthermore, some theoretical conditions exist under which mutation accumulation may not increase genetic variances (Charlesworth & Hughes, 1996). Overall, the study of genetic variances as a function of age has not been particularly clarifying.

The other quantitative genetic test might seem to be more promising: searching for negative genetic correlations between life-history traits at early and late ages as evidence for antagonistic pleiotropy. Such negative correlations have sometimes been found (e.g., Rose & Charlesworth, 1981a), but in many cases they have not (Rose et al., 1996). There are two experimental artifacts that tend to bias genetic correlations toward positive values: inbreeding depression and genotype-by-environment interactions (Rose, 1991). Inbreeding depression occurs when populations have been inbred to such an extent that most or all life-history traits are depressed.

Under these conditions, it is expected that populations that vary in their degree of inbreeding depression will show positive covariation among their life-history traits. Genotype-by-environment interactions tend to generate positive covariation among life-history traits when populations are studied under conditions to which they are not well-adapted, such as the laboratory conditions imposed on organisms only recently brought into captivity. For these reasons, quantitative genetic estimates of genetic correlations provide unreliable tests of patterns of pleiotropy between early and later life-history traits.

Michael Rose and collaborators have spent considerable time and effort on quantitative-genetic tests of evolutionary mechanisms for aging. But looking back, this approach appears to be unsatisfactory for testing the evolutionary theory of aging.

2.3.4. *Experimental population genetics of aging*

If quantitative genetics research on aging has been stymied by inbreeding depression and genotype-by-environment interactions, then the material of experimental population genetics makes such challenges inevitable. That is because the standard protocols of this approach start with inbred samples derived from wild populations.

When such samples are inbred in captivity, they lack the opportunity to adapt to laboratory conditions; in fact, that has been the intent of such inbreeding. Thus, the standard inbred lines of population genetics research will consistently have depressed life-history traits, often including accelerated aging. Furthermore, the impact of inbreeding will vary *among* inbred lines, producing a spurious positive covariation when inbred lines are used to test for patterns of genetic correlation between early and late ages. This approach undermines the utility of such material for the study of antagonistic pleiotropy in the evolution of aging.

Crossing inbred lines leads to additional complications because of the potential for *epistatic interactions* across the different inbred lines. (Epistasis refers to nonadditive interactions among alleles at different loci in the determination of phenotypes.) Fundamentally, there is no simple way to cross inbred lines to determine the evolutionary genetic mechanisms that underpin the evolution of aging, among other functional traits, in the original wild population from which the inbred lines were derived.

2.3.5. *Experimental evolution of aging*

Empirical research on the evolution of aging made compelling progress only after the field turned to experimental evolution. This research began with unreplicated pilot projects (e.g., Rose & Charlesworth, 1980), but it soon featured extensive replication across multiple experiments and multiple laboratories (e.g., Luckinbill et al., 1984; Rose et al., 2004).

Experimental evolution has provided rigorous tests of the key role of natural selection in determining both the onset and the cessation of aging. Theoretically, Hamilton's (1966) analysis of the forces of natural selection implies that aging should begin only after the age of first reproduction in a population's life history. Later theoretical work also indicates that aging may stop after the last age of reproduction in the population's life history (Mueller et al., 2011). The first of these predictions has been tested repeatedly by delaying the age of first reproduction in replicated outbred populations of *Drosophila* (e.g., Burke et al., 2016; Rose, 1984b), with corroborative results obtained consistently (see Figure 2.2). The second of these predictions has also been tested by comparing outbred *Drosophila* populations with different last ages of reproduction (Rose et al., 2002), again with

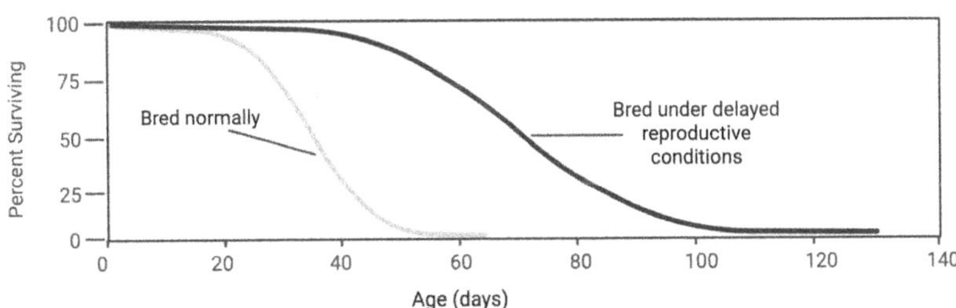

Figure 2.2: Delaying reproduction in evolving populations of fruit flies leads to increased adult survival and delayed senescence. Hamilton (1966) predicted that aging should begin only after the age of first reproduction in a population's life history. This prediction has been tested repeatedly by delaying the age of first reproduction in replicated outbred populations of *Drosophila*, with definitively corroborative results obtained in many experiments.

corroborative results. (More details are provided in Chapter 7.) These are some of the most direct experimental tests of a specific formal prediction in evolutionary biology as a whole.

Outbred *Drosophila* populations that have evolved different patterns of aging in the laboratory have also provided useful material for the study of the evolutionary genetic mechanisms of aging. In particular, a common result has been that the experimental evolution of postponed aging is associated with reduced early fecundity and slowed larval development (Rose et al., 2004), a pattern expected with antagonistic pleiotropy. These results are sometimes impacted by genotype-by-environment interactions, resulting in inconsistencies across experiments that may reflect slight differences in procedures to measure fecundity (see Leroi, Chippindale & Rose, 1994). Nonetheless, the results from experimental evolution have supported the role of antagonistic pleiotropy in the evolution of aging.

It has been more difficult to use experimentally evolved material to test the possible importance of mutation accumulation. However, one such test that has been performed involved crossing experimentally evolved replicate populations to test for hybrid vigor among life-history traits at later ages. Such hybrid vigor would imply that mutation accumulation had led to idiosyncratic degradations of function in each inbred line. Evidence for such hybrid vigor has been found in particularly small populations denied the opportunity to reproduce at later ages (e.g., Mueller, 1987). However, it was not observed in other experimental tests using populations maintained at larger sizes (e.g., Rose et al., 2002). This difference may indicate that mutation accumulation is more important to the evolution of aging in small populations, because genetic drift is more pronounced relative to selection in that case.

2.3.6. *Overview of various methods in the evolutionary biology of aging*

Comparative biology has made a major contribution to evaluating the evolutionary theory of aging by showing that aging evolves whenever the force of natural selection declines during an organism's reproductive adulthood. Otherwise, however, this approach suffers from confounding factors.

Quantitative genetics has provided suggestive results with respect to evaluating the evolutionary genetic basis of aging. In particular, it has sometimes, but not always, found negative genetic correlations between early fecundity and later adult survival (Arnold & Rose, 2023; Rose et al., 1996).

The heavily inbred material of experimental population genetics is not generally useful for testing the evolutionary theory of aging. Indeed, such material is poor for evaluating most hypotheses concerning the functional evolution of wild populations, except for the mere demonstration that such populations typically have heritable variation for most such traits.

Experimental evolution has provided the most useful approach for testing and refining the evolutionary theory of aging, especially across the dozens of experiments that have used experimentally evolved *Drosophila* populations (see Arnold & Rose, 2023; Rose et al., 2004). Thus, the *why* of experimental evolution is because it is often the most powerful method for making progress in tackling the big questions of evolutionary biology.

Furthermore, differentiated populations generated by experimental evolution constitute material that is useful for investigating the functional (i.e., physiological, morphological, and behavioral) basis of aging and other traits of interest (e.g., Barter et al., 2019; Graves et al., 1988; Rose et al., 2004). As such, we contend that experimental evolution is a powerful tool for use across a wide range of biological research.

Chapter 3

The How of Experimental Evolution

In the previous chapter, we explained that *experimental evolution* can be broadly defined as conducting real-time evolution experiments under well-defined and reproducible conditions, which thereby permit powerful inferences concerning the joint action of natural selection, genetic drift, mutation, and recombination. We introduced four essential features of experimental evolution: replication, controls, genetic variation, and observing change across many generations. We will now discuss each of these requirements in greater detail. We will then present issues of experimental design that are specific to two broad categories of organisms widely used in evolution experiments: microbial and nonmicrobial.

3.1. How to Run an Evolution Experiment

3.1.1. *Replication*

In general, experimental evolution requires multiple populations evolving under identical selective regimes. Perhaps the best-known example of an evolution experiment using a microbe is the Long-Term Evolution Experiment, or LTEE for short. The LTEE has 12 replicate populations, all of which were founded from one strain of the bacterium *Escherichia coli* (Lenski et al., 1991). The choice of 12 replicates was based on some practical limitations regarding the culture vessels, equipment, and labor available when the LTEE began in 1988. However, it is now often feasible to have greater replication. For example, some recent evolution experiments with microbes have used microtiter plates and robotic transfers to

25

enable 96-fold replication of the evolving populations (Johnson et al., 2021; Lang et al., 2013; Lenski, 2023).

Having more replicate populations increases the statistical power to address the big questions we presented in Chapter 1. However, the amount of replication that is feasible depends on the organisms under study. Larger organisms, such as fruit flies, require larger vessels (vials, bottles, or cages) for housing populations of sufficient size. Other species may pose even greater constraints; for example, small mammals such as mice not only require more space but also raise considerations related to social behaviors, parental care, and animal welfare. In a study with mice, Garland et al. (2002) had four replicate lines for mice that were selected for high activity along with four control lines. The space demands and other costs of evolution experiments with larger organisms have meant that few such experiments have been sustained for many years. As a consequence, evolution experiments are usually conducted using smaller species such as microbes, nematodes, insects, small fish (e.g., guppies), small mammals (e.g., mice), and small plants (e.g., *Arabidopsis*). The more successful of these experiments use a minimum of four replicate treatment populations, with a similar number of control populations.

3.1.2. *Controls*

In experimental evolution, the most important controls are populations that are *not* evolving in and adapting to the new selective regime that is the focus of the study. The best controls are essentially identical to the ancestral population from which the experimental populations were derived. As explained in Chapter 2, samples of the founding ancestral population can be stored frozen for many microbes and then later revived. The same is feasible in some multicellular organisms, like *Caenorhabditis elegans*, a nematode widely used as a model organism in many fields of biological investigation. These control populations have been called a "frozen fossil record" (Lenski & Travisano, 1994). This type of control allows derived populations to be directly compared with thawed samples of the ancestors. The frequency with which samples are frozen may depend on the anticipated rate of evolution as well as freezer capacity. During the first year of the LTEE, samples were frozen after every 15th daily transfer, corresponding to 100 bacterial generations; as the experiment progressed, later samples were frozen at 500-generation intervals (i.e., every 75th transfer). In principle, genotypes may differ in their tolerance to freezing and thawing; in practice, however, such effects tend to be minor and should not systematically bias results with respect to most general hypotheses (Lenski, 2023).

In evolution experiments with *Drosophila*, it is a common practice to start populations by producing an outbred population from crossing many isofemale lines (e.g., Kellermann et al., 2015; Tobler et al., 2015). The same procedure is sometimes used to produce controls at the relevant generations of the experiment by crossing the same inbred lines that are maintained in parallel with the outbred treatment populations. This procedure is meant to reconstitute the ancestors of the experimental populations. It assumes that the ancestral inbred lines are fully homozygous so that they do not evolve, or evolve minimally, over the duration of the experiment (see Nouhaud et al., 2016, for a test of this assumption). With homozygous lines, however, an accidental, rare, immigrant fly could undermine their value. This approach thus requires a high level of care in stock maintenance that is only rarely achieved in *Drosophila* laboratories. Other important limitations of this approach will be discussed in a later chapter.

The second-best kind of controls are outbred populations derived from the same ancestral population as the treatment populations, which are maintained under culture conditions where the control populations are not expected to evolve and adapt much, if at all. These controls are most useful if the ancestral population had already achieved approximate evolutionary stasis in the culture regime that was imposed prior to the start of the experiment. The time required to achieve such stasis is likely to be at least 50 generations, based on the results of Matos et al. (see Chapter 7). As a rule of thumb, we suggest that 100 generations in a consistently maintained culture regime *prior to* the start of an evolution experiment should suffice for relatively outbred populations maintained in the laboratory. [See also Burke et al. (2016) for more data on this point.] In studies where the focus is specifically on adaptation to the lab environment, some other type of control becomes necessary. For example, one might use other populations that have already adapted to the lab for many generations. These control populations should ideally be derived from the same or similar natural populations as the new populations undergoing evolution in the lab (e.g., Matos et al., 2002; Simões et al., 2007, 2017; see also Chapter 7).

3.1.3. *Genetic variation*

In the previous chapter, we explained that adaptation by natural selection requires genetic variation. Experimental microbial populations are typically founded from a single colony that is the outgrowth of a single haploid cell. However, a single colony of bacteria typically has many millions of cells, and bacterial populations propagated in liquid cultures often have billions of individuals, depending on the culture volume and growth medium provided. In microbial evolution experiments, new mutations are typically the sole source of genetic variation, and they provide ample scope for adaptation given such large populations.

Mutation rates can vary widely depending on the particular microbe. Among the factors that influence the genome-wide mutation rate are the size of the genome and the nature of the genetic machinery. Most bacteria have genome lengths between 1 and 10 Mb (1 Mb = 1 million base pairs). For example, the *E. coli* strain used to start the LTEE has a 4.6-Mb genome. The brewer's yeast *Saccharomyces cerevisiae* is now widely used in evolution experiments, and it has a genome of ~12 Mb. Viruses have also been used in some evolution experiments; although they tend to be smaller, there is a tremendous range from a few Kb (1 Kb = 1 thousand base pairs) to over a Mb. The rate of mutation also depends on the genetic machinery. Mutation rates are much higher in viruses that have RNA-based genomes than in DNA viruses. Most cellular organisms have proofreading mechanisms that reduce mutation rates below what would otherwise occur. However, some bacterial strains have defects in those mechanisms, which can cause mutation rates to increase by 100-fold or so. As a point of reference, the ancestral *E. coli* in the LTEE has a rate of point mutations of $\sim 2 \times 10^{-10}$ per base pair (Tenaillon et al., 2016); with a 4.6-Mb genome, that means roughly one new point mutation is produced in every 1,000 cell replications. Each day, there are about half a billion cells produced in each LTEE population, which means there are roughly half a million point mutations generated every day in each population. There are also various insertions, deletions, and other types of mutations.

Thus, mutations alone provide a substantial, continuous supply of genetic variation in microbial populations.

In addition, parasexual processes like plasmid-mediated conjugation and virus-mediated transduction can introduce entirely new genes from the same or other species, thus providing larger blocks of genetic variation than mutations provide. In nature, these parasexual processes are fairly common among bacteria, and they also sometimes occur in microbial eukaryotes like yeast. Therefore, it is important to determine in advance whether these mechanisms are operating in an experimental system. The ancestral strain of the LTEE lacks plasmids and functional viruses; it was chosen so that *de novo* mutations would provide the only source of genetic variation (Lenski, 2023). It is also important to recognize that the parasexual processes of conjugation and transduction involve genetic vectors (plasmids and viruses, respectively) that themselves can evolve, which introduces an additional complication, although the resulting coevolution may well be of scientific interest. In any case, it is important to know whether parasexual processes are occurring and to take them into account when analyzing an evolution experiment. For example, one could sequence the vector genome and analyze the phenotypic traits (e.g., transmission rate) that are relevant to its interaction with the host microorganism.

With diploid sexual organisms, experimental populations typically begin with abundant standing genetic variation, so long as the populations are not highly inbred (and thus mostly homozygous). This initial variation allows a rapid response to selection. Moreover, Mendelian recombination can regenerate abundant genetic variation by exposing new combinations of existing alleles, in addition to new variation produced by mutations. The existence of such abundant genetic variation makes it especially important to sustain a consistent selection regime for the evolving outbred control populations, which are the controls typically used in evolution experiments with *Drosophila*.

3.1.4. *Observing change across many generations*

Experimental evolution fundamentally depends on allowing sufficient time for evolution to occur. It is often more useful to think in terms of the number of generations over which experimental populations are subjected to novel selection regimes, as opposed to standard time units like days and years. The number of generations can be readily estimated when the experimental protocol has discrete (i.e., nonoverlapping) generations. There are also formal methods for estimating the number of generations in the context of age-structured populations with overlapping generations (Charlesworth, 1994). In any case, an experiment with many generations allows the effects of natural selection, mutation, recombination, and genetic drift to accumulate, resulting in greater differentiation of experimental populations relative to their ancestors (or controls) and relative to each other. The ongoing LTEE (discussed in detail in Chapter 6) has set the gold standard for evolutionary time in an experiment with microbes, as its 12 *E. coli* populations have surpassed 80,000 generations (Lenski, 2023). The longest sustained evolution experiments with *Drosophila* are from the Rose laboratory, with some populations reaching ~1,000 generations before 2020 (Graves et al., 2017; Phillips et al., 2016; see Chapter 7).

These long-sustained experimental populations have undergone substantial phenotypic and genetic differentiation from their ancestors (in the LTEE) and controls (in the *Drosophila* work). Such differentiation, when combined with replication of experimental populations, strengthens the statistical inferences concerning their evolution. Patterns of differentiation *within* groups of populations that share a common selection history can also be combined with patterns of differentiation *among* populations with different histories in a variety of statistical analyses (Blount et al., 2018; Deatherage et al., 2017; Rose et al., 2004). These analyses grounded in replicated, long-running evolution experiments are generally more powerful than those based on experimental population genetics or quantitative genetics, as discussed in Chapter 2. In effect, long-running evolution experiments amplify genetic signals, allowing one to infer the aggregate or singular effects of the alleles that underpin evolutionary processes. That inferential power is the fundamental reason why we and many others have increasingly turned to experimental evolution to answer some of the big questions in evolutionary biology, even in biology as a whole.

3.2. Running an Evolution Experiment with Microbes

Evolution experiments using microorganisms exemplify some of the best practices of the field. Nonetheless, there are important design issues that should be carefully considered that arise from particular features of micro-bial systems. These issues include culturing procedure, growth medium, founding strain, cross-contamination, and population size.

3.2.1. *Culturing procedure*

There are three culture regimes that have been used in evolution experiments with microbes: serial transfer, continuous culture, and single-colony passaging (Barrick & Lenski, 2013).

Serial transfer works by sampling a specified portion of a culture (e.g., 1%) at the end of a fixed period of time (e.g., 24 h) in a closed vessel (e.g., flask) and then transferring that sample to a sterile vessel containing fresh growth medium (Figure 3.1, panel a). This method provides opportunities to control both the ecology and evolution of experimental populations, at the cost of some manual labor. (In some labs, robotic systems are now used to perform the transfers, but they still require considerable oversight.) In particular, one can set the final population size by varying the concentration of the limiting resource in the growth medium or by varying the volume of the culture. One can also set the minimum "bottleneck" population size and thereby modulate the influence of random genetic drift by varying the dilution factor (Izutsu et al., 2024).

Continuous culture typically occurs in devices called chemostats (Figure 3.1, panel b). A chemostat is a vessel containing a population of cells that is sustained by the continuous inflow of medium together with an equal outflow of medium. Importantly, the outflow includes the cells present in that medium. The balance between inflow and outflow thus provides a stable population size and a constant level of resources in the medium — a level that is just sufficient to balance the growth rate of the cells with the rate at which they are washed out of the chemostat. Chemostat cultures are elegant for mathematical models and physiological studies because of the constancy of the population size and ecological conditions (Kubitschek, 1970). In practice, however, chemostats are challenging owing to the expertise necessary to run and maintain them as well as the expense of customized equipment and materials. Cells may also stick to the walls of the culture vessel. When that occurs, it violates a key assumption of a chemostat, namely that cells are removed at the same rate

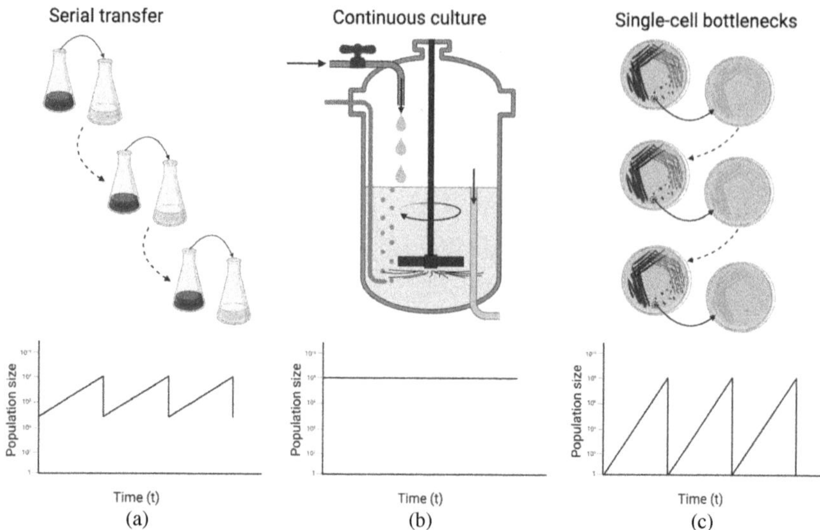

Figure 3.1: Three protocols for propagating microbes in evolution experiments. (a) Serial transfer: A population can be propagated by transferring a defined portion of the population to fresh medium at regular intervals. The population size will fluctuate between minimum and maximum values. (b) Continuous culture: A population can be propagated in a chemostat or similar device, with sterile medium flowing in at a constant rate and spent medium and cells flowing out at the same rate. The population will maintain a constant population size. (c) Single-cell bottlenecks: A culture can be streaked onto an agar plate to yield isolated colonies, each of which derives from a single cell, with this process repeated at regular intervals. The population size will fluctuate between 1 and the number of cells in a colony, with genetic variation purged at each single-cell bottleneck (adapted from Figure 1 in Barrick and Lenski [2013]).

as the outflow of the medium. Moreover, in an evolution experiment, one may inadvertently select for mutants that have an increased propensity for sticking to the chemostat vessel. Finally, owing to the continuous flow of medium and gases in the space above the liquid culture, chemostats are prone to contamination by environmental microbes. Owing to these practical challenges, chemostats are not used in most microbial evolution experiments, at least not those intended to run for a long duration.

With single-colony passages, a microbial population is diluted and spread over a Petri dish containing agar and nutrients that allow cells to grow into colonies that are visible to the naked eye (Figure 3.1, panel c).

Each colony contains millions of cells but is the outgrowth of a single cell. By passaging a cell line from one colony to the next, one can repeatedly impose single-cell bottlenecks that purge all of the genetic variation from the lineage. Natural selection cannot operate without genetic variation, and so this approach prevents adaptation by natural selection. Nonetheless, mutations still occur, and this method (especially when coupled with whole-genome sequencing) offers a way to estimate the underlying mutation rate by minimizing the complicating effects of natural selection on the rate at which mutations accumulate over time (Long et al., 2018; Tenaillon et al., 2016). In short, this mode of microbial propagation is useful for isolating and studying one evolutionary process, namely mutation, to the exclusion of natural selection.

3.2.2. *Growth medium*

The choice of culture medium is an important part of the design of any experiment with microbes. Attention must be paid to standardizing the medium and other conditions (e.g., temperature) across treatment and control populations. The use of chemically defined media can be helpful in this regard. Whether an evolution experiment is explicitly focused on the culture medium or on some other selective agent (e.g., the presence or absence of a predator), adaptation to the medium is likely to occur or, at least, is a possibility that must be considered. Thus, genetic variants that increase in frequency in the evolving populations may reflect such medium-specific adaptation, even if it is not the main objective of the experiment.

3.2.3. *Founding strain*

The ancestral strain chosen to begin a microbial evolution experiment is another important consideration. A common choice is to use an ancestral strain that has already been fully sequenced, although whole-genome sequencing has become sufficiently fast and inexpensive that this issue is less important now than in the past. In any case, it is a good practice to sequence the material you receive from a biological supply company or collaborator, because mutations may have arisen that are not reported in the reference genome (Graves et al., 2015). A related point is that a potential founding strain might also be one that has been used as a model system in other fields, such as molecular biology and physiology; if so, there might be better genetic tools or more functional information that could be useful when analyzing the results of an evolution experiment. Regardless of the choice of the ancestral strain, it is usually advisable to start each population from a different single colony that represents the outgrowth of a single haploid cell. By doing so, one ensures that each replicate population has an independent history of mutations (Lenski, 2023). Therefore, if similar or even identical changes occur in the replicate populations, one can be confident that those changes reflect parallel evolution (Lenski, 2017a).

3.2.4. *Cross-contamination*

In the context of an evolution experiment, cross-contamination refers to the inadvertent mixing of organisms between replicate populations. This possibility is different from a contaminant of a different species, which should be easily detected based on phenotypic or genetic characteristics. If not detected, cross-contamination compromises the independence of the replicate populations, because a beneficial mutation that appears in one population could then spread through other populations. One way to monitor for cross-contamination is to use two or more variants of a founding strain that can be distinguished based on some easily recognized trait. One would then alternate the variants when handling the populations, such that any inadvertent cross-contamination would typically introduce a variant that differed from what was expected. This approach has been used throughout the LTEE; six populations founded by arabinose-minus (Ara$^-$) and six by arabinose-positive (Ara$^+$) variants of the same ancestral strain have been strictly alternated during the daily serial transfers (Lenski, 2023). The arabinose marker causes the bacteria to make different-colored colonies on a particular agar medium, and the researchers periodically check for that phenotype. If a cross-contaminant is detected, one can go back to the most recent frozen sample and restart the affected populations from that timepoint. Nowadays, genome sequencing allows a further check, as each LTEE population has fixed many mutations that distinguish it from any of the other lineages (Tenaillon et al., 2016).

Indeed, identical alleles that rise to high frequency in two or more replicate populations may suggest cross-contamination, as opposed to truly parallel evolution. For example, Tajkarimi et al. (2017) selected *E. coli* for ionic silver resistance. They sequenced samples from 19 populations, and 13 had a mutation in a gene, *cusS*, that encodes a histidine kinase involved in sensing copper and silver. Moreover, 10 of the *cusS* mutations were identical. This mutation was shown to be superior at binding to silver when compared to other mutations that evolved in that experiment. However, it is unclear whether the identical alleles resulted from mutations that occurred independently in the 10 populations or, alternatively, the identical alleles derive from a single mutation that spread through many replicates after inadvertent cross-contamination.

3.2.5. *Population size*

The number of individuals in each population is an important consideration in designing any evolution experiment. Microbial populations tend to be large, even when grown in small volumes, and so one might not give the population size too much thought. In fact, however, it deserves attention. For example, the LTEE experiment was designed to keep the bacterial densities a bit lower than most microbiologists would have done by using a low concentration of glucose, which is the growth-limiting resource in the medium (Lenski et al., 1991). This lower-than-maximum density was used to limit the accumulation of metabolic byproducts in the cultures, as some byproducts can provide secondary resources or act as growth inhibitors. While such effects can be interesting, they also complicate the analysis of adaptation by promoting frequency-dependent selection (Rozen & Lenski, 2000).

At the same time, the LTEE was designed to avoid severe population bottlenecks, which, as noted earlier, reduce the genetic variation that fuels adaptation by natural selection. In the LTEE, the serial transfer regime involves transferring 0.1 mL of the previous day's culture into 9.9 mL of fresh medium each day. The bacteria reproduce by binary fission, and therefore the 100-fold dilution and regrowth lead to ~6.6 generations between successive transfers ($\log_2 100 \approx 6.6$). The LTEE could have produced more generations by using a more extreme dilution, but this was not done in order to avoid more severe bottlenecks. Given the size of each LTEE population before the daily transfer (~5 × 10^7 cells/mL × 10 mL), the minimum (bottleneck) size is ~5 million cells, providing ample opportunity for genetic variation to persist across transfers. Though it was not known when the LTEE started, recent analyses indicate that the 100-fold dilution factor was close to the optimum for maximizing the rate of adaptation given the other parameters relevant in this system (Izutsu et al., 2024).

3.3. Running an Evolution Experiment with Larger Organisms

While the preceding section focused on best practices for designing evolution experiments that use asexual microbes, the same overarching principles apply to sexual multicellular organisms. However, some of the specific techniques do not translate directly from microbial experiments. For example, cryopreservation does not usually work for most macroscopic organisms, so it is generally impossible to perform direct comparisons between evolved and thawed ancestral stocks. (However, see Teotónio et al. [2017] for the use of cryopreservation in experimental evolution with nematodes.) Given these technical issues, we will proceed through the same design issues of physical procedures, nutrition, founders, cross-contamination, and population size for evolution experiments with macroscopic species.

3.3.1. *Physical procedures*

Evolution experiments that use animals or plants must adapt their procedures to the idiosyncratic biology of the species involved. Aquatic species may face the same challenges associated with contamination that arise when using chemostats for research with microbes. Some terrestrial species can fly, making the design of effective enclosures more difficult. Cultures of fruit flies include multiple life stages: larvae that crawl through a semi-liquid medium, pupae that typically attach to the sides of the container, and adults that can both walk and fly (see Figure 3.2). Plant

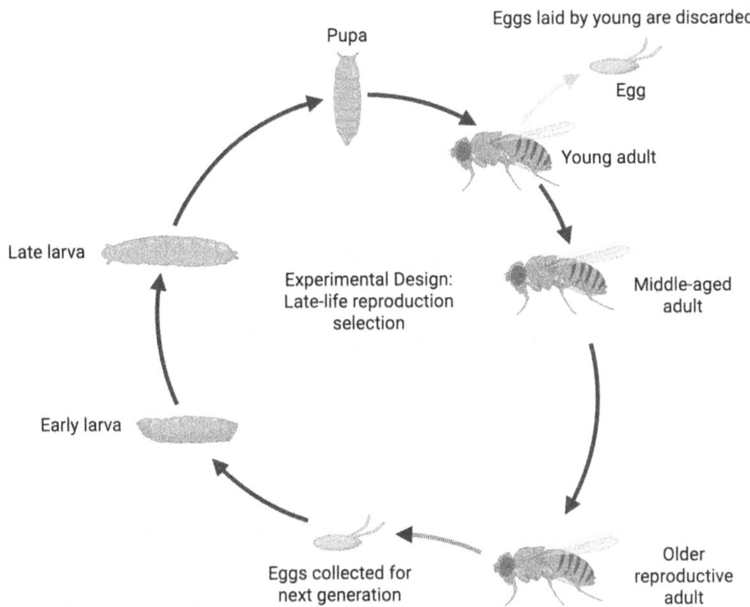

Figure 3.2: Life cycle of fruit flies, illustrating how the evolution of delayed senescence can be promoted by using the eggs produced by older adults while discarding eggs from young adults. Fruit fly cultures include multiple life stages: larvae that crawl through a semi-liquid medium, pupae that typically attach to the sides of the container, and adults that can both walk and fly.

species will typically require light for photosynthesis, but they are immobile, which is convenient.

3.3.2. *Nutrition*

Animals, of course, require food. A common mistake is to supply experimental populations with arbitrary foods at the convenience of the experimenter. The problem with this approach is that adaptation to particular sources of nutrition is a common feature of populations evolving in the laboratory (see Rutledge et al., 2020). Thus, maintaining the same dietary regime as that given to the recent ancestors of the experimental populations is among the most important principles of good experimental design with animal species. Similar considerations would apply for plants with respect to light, water, soil, and other inputs.

3.3.3. *Founders*

One of the most important issues when working with animals and plants is the genetic characteristics of the founding population. First and foremost, populations started from few founders will likely have lost most of the genetic variation of the source population due to the severe bottleneck. Moreover, evolution experiments work best when starting with a founder population that is outbred. Inbred ancestors will typically have lost allelic variation, including variants sustained by balancing selection in large, outbred populations from nature. Besides the depressed fitness that is typical of inbred populations, such populations will also have elevated and idiosyncratic patterns of homozygosity. Allelic homozygosity across most loci is an inherent barrier to evolutionary change, especially in populations that rely on standing genetic variation to evolve (e.g., Burke et al., 2010), as is typically the case in evolution experiments with macroscopic species. Populations derived from highly polymorphic founders allow more loci to contribute to the response to selection, thus shortening the time necessary for an experiment to produce meaningful results (Burke, 2023).

3.3.4. *Cross-contamination*

The risks of cross-contamination depend on the biology of the macroscopic species and the physical methods used to contain the study populations. Plants cannot move and might seem ideal with respect to minimizing the potential for cross-contamination. However, their pollen is tiny and readily dispersed by air movement or pollinators, leading to a risk of cross-contamination. Mice are unlikely to escape from one cage and then get into another, but flying insects like *Drosophila* can move between enclosures. In an evolution experiment that starts from outbred founders and sustains large populations, an occasional migrant is unlikely to severely impact the outcomes (in contrast to microbes, where a single contaminating cell with a fitness advantage could profoundly alter the evolutionary trajectory of the affected population).

3.3.5. *Population size*

Maintaining reasonably large populations during an evolution experiment with any sexual species is crucial to avoid inbreeding. For organisms such as fruit flies, the minimum population size should be on the order of a thousand individuals. Some evolution experiments with *Drosophila melanogaster* in enclosures have even started with populations as high as 100,000 flies (Rudman et al., 2022). By contrast, evolution experiments with mice often have populations ranging from about 10 to 100 breeding adults (Firman & Simmons, 2011).

Another advantage of large population sizes is that evolution in diploid sexual organisms is driven primarily by standing genetic variation (Burke, 2012). Larger starting populations harbor more genetic variation. Furthermore, whether rare beneficial alleles can become common will depend on both the strength selection for those alleles and the number of generations. Thus, evolution experiments with multicellular sexual organisms should begin with populations as large as feasible and run for as many generations as possible. Even when the population starts with genetically diverse founders, that diversity can only be sustained if the populations remain large throughout the experiment. The loss of diversity and the adverse effect of inbreeding are two of the biggest problems facing evolution experiments that use macroscopic organisms.

3.4. Types of Selection Regime

Evolution experiments can employ various kinds of selection, among them domestication in the laboratory, adaptation to one or more specific environments or demographic regimes, and reverse selection. We consider each in turn.

3.4.1. *Domestication*

The domestication of various organisms is humanity's oldest kind of evolution experiment, even if the underlying process was not understood for many millennia. Among the successfully domesticated organisms are not only dogs and livestock but also the agricultural crops that sustain us and even the yeast we use in baking and brewing. Darwin began *The Origin of Species* (1859) with a chapter on domestication as a way to convince readers of the power of selection to change and reshape organisms. In the context of experimental evolution, domestication refers to the sampling of organisms from the wild, which are then propagated in and adapted to laboratory conditions. The Matos laboratory has focused on this process extensively over the last 30 years, chiefly using *Drosophila subobscura*. That work will be discussed in detail in Chapter 7.

3.4.2. *Adaptation to specific environments or demographic regimes*

Even after a population has been propagated in and adapted to generic laboratory conditions, a common approach in experimental evolution is to change a particular aspect of the culture conditions. Such selection might involve a change in temperature (Bennett et al., 1992; Kellermann et al., 2015; Santos et al., 2023; Tobler et al., 2015), resources (Travisano et al., 1995), the age-specific opportunity for reproduction (Rose et al., 2004), or the dilution regime (Izutsu et al., 2024). Selection in such experiments can occur in several different ways. First, it may involve the extermination of those individual organisms that fail to meet a particular criterion based on their performance; this process is sometimes referred to as "laboratory culling" (Rose et al., 1990). Second, it may occur by "artificial selection" (Falconer, 1960). Here, individuals are selected based on some measurable trait, and efforts are made to forestall all natural selection other than the selection based on the chosen criterion. An artificial-selection protocol requires that the trait of interest be measured for all individuals in each generation, with those measurements used to identify the individuals that will serve as the parents for the next generation. The protocol typically seeks to equalize the number of offspring among selected individuals and, unlike culling, avoid selection that might act on any other traits. Last, but not least, is what has been called "natural selection in the laboratory" (Feder et al., 2000; Gibbs, 1999; Rose et al., 1990). In these experiments, there is no explicit culling or artificial selection based on any particular trait. Instead, natural selection — that is, differential survival and reproductive success based on any and all traits — is allowed to proceed exactly as it does in nature, except in the context of a constructed laboratory environment.

3.4.3. *Reverse selection*

As its name suggests, reverse selection is done by starting with populations that have previously been subjected to "forward" selection for a particular trait, or in a particular environment, using any of the selection schemes described in the previous section. Some reverse-selection experiments simply stop imposing the forward-selection regime, returning the experimental populations to the original culture conditions prior to the initiation of that experiment. These are sometimes called "relaxed selection" experiments. Other reverse-selection experiments actively impose selection on experimental populations using criteria that are the opposite of the previous forward-selection conditions. For example, the ancestral environment might have had an ambient temperature of 21°C, with forward selection for adaptation to 25°C. The reverse selection could then be "relaxed" by returning the 25°C populations to 21°C. Alternatively, the reverse selection could be active, with the 25°C populations subjected to an even lower temperature of 18°C (see Teotónio and Rose [2000] for further discussion of reverse selection and evolutionary reversibility).

3.5. Chapter Summary

Population replication, controls, genetic variation, and evolutionary time are all essential aspects of the design of an evolution experiment. A number of experimental procedures depend on the type of organism that is used to test the relevant evolutionary hypotheses, with important differences between microbial and macroscopic species. Evolution experiments may also involve several distinct types of selection, including domestication, adaptation to specific environments or demographic regimes, and reverse selection. The way in which selection is imposed on populations can also vary from culling and artificial selection to allowing natural selection to proceed in the laboratory as it would in nature.

Chapter 4

Phenotypic Responses to Experimental Evolution

The historical antecedents of experimental evolution were selection experiments conducted in the first half of the 20th century. Most of these experiments sought to enhance agricultural productivity, from corn oil content to egg-laying by hens. Only a few selection experiments were done to test basic scientific ideas before about 1950 (e.g., Castle & Phillips, 1914).

From 1950 to 1980, more evolutionary biologists began to probe some of the big questions of our field using selection experiments. However, their technique was often unsatisfactory, especially with respect to replication and the use of appropriate controls (reviewed in Wright, 1977).

Starting in the 1980s, experimental evolution became a commonly employed research strategy in evolutionary biology. While the elements of good experimental design (described in Chapter 3) were soon widely appreciated, the challenges of appropriate phenotypic assays took some time to be met.

Here we use a basic Darwinian hierarchy to organize our discussion of phenotypic traits and their measurement in the context of experimental evolution. We begin with direct assays of Darwinian fitness, followed by the life-history traits and demographic parameters that are components of fitness. We then discuss organism-level traits that reflect morphology, physiology, and behavior. Finally, we discuss biochemical metabolites and processes that are intermediaries between the genome and organismal functions.

4.1. Darwinian Fitness

The value of directly assaying Darwinian fitness in asexual microbes contrasts with the difficulty of doing so in other systems for experimental evolution. Therefore, we will discuss the two situations and challenges separately.

4.1.1. *Asexual microbes*

For evolution experiments conducted with asexual microbes, the measurement of Darwinian fitness (i.e., the overall fitness of the whole organism, often expressed relative to some standard) is fairly straightforward. Such projects usually feature frozen ancestral stocks, as well as genetic markers that can be used to differentiate those ancestors from the experimentally derived microbes (Lenski, 2023; Lenski et al., 1991). In brief, the assay of fitness proceeds by first separately acclimating the ancestral and derived samples to the culture conditions in which they will compete. They are then mixed together and enumerated at the start and finish of the assay based on the genetic marker (see Figure 4.1). The fitness of the experimentally evolved microbe can then be expressed as the ratio of its realized (i.e., net) growth rate relative to that of the ancestral type. This approach is exemplified by the fitness assays performed in the LTEE, using derived and ancestral *Escherichia coli* that are marked with the alternative versions of the arabinose-utilization gene.

Fitness assays are typically performed in the same medium and other conditions, and over the course of the same culture cycle, as used in the evolution experiment itself. However, the fitness assays are also sometimes run under other conditions, such as at different temperatures or with alternative resources (Bennett & Lenski, 1993; Travisano & Lenski, 1996). In that case, the relative fitness values measured under these alternative conditions are called *correlated* responses in order to distinguish them from the *direct* response to selection in the environment where the derived forms evolved.

In principle, this experimental strategy does not require the use of single-celled microbes. The nematode *Caenorhabditis elegans* can also be frozen, rarely has sex, and has an abundance of candidate marker alleles. Thus, its fitness can be assayed in competition using similar methods to those used with microbes (Walker et al., 2000).

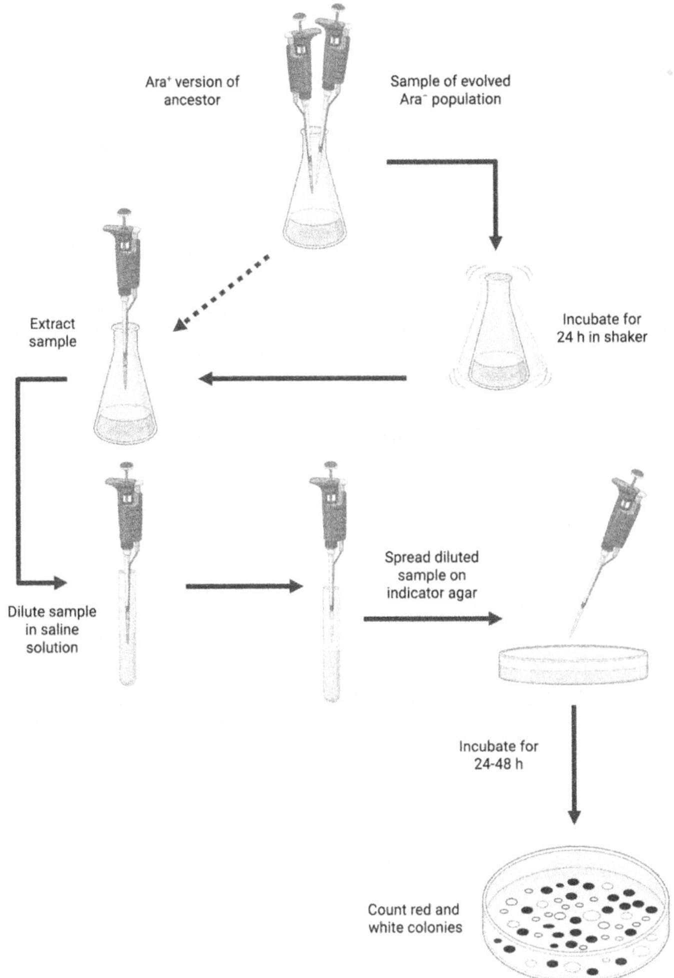

Figure 4.1: Competition assays used to estimate relative fitness in the LTEE. A sample of an evolved population is mixed with the ancestral strain that has the opposite arabinose-utilization marker state (in this case Ara⁻ and Ara⁺, respectively). This marker is selectively neutral in the LTEE environment. An initial sample of the mixed culture is diluted in saline solution, and an aliquot is plated on an indicator agar that allows Ara⁻ and Ara⁺ cells to be distinguished because they make red and white colonies, respectively, after incubation. Meanwhile, the mixed culture is incubated for 24 h, after which a final sample is diluted in saline solution and an aliquot plated on the indicator agar. After adjusting for the initial and final dilution factors, the growth rate of each competitor can be estimated. Relative fitness is then calculated as the ratio of the evolved population's growth rate to the ancestor's growth rate during the head-to-head competition assay. Barrick et al. (2023) illustrate and describe these procedures in greater detail.

4.1.2. *Sexually reproducing organisms*

Once sexual reproduction enters the picture, it is more difficult to accurately estimate Darwinian fitness by performing direct competitions between genetically marked ancestral and derived samples. The difficulty occurs because sexual recombination will dissociate any genetic marker from the fitness effects caused by the allelic differences between the ancestral and experimentally evolved organisms. Nonetheless, a few studies have performed competitive fitness assays with some success, while acknowledging the resulting limitations (e.g., Teotónio et al., 2002).

Given the problem of estimating relative fitness in sexual organisms using competition assays, there are other approaches for estimating fitness in absolute terms. One approach is to measure the total productivity of a cohort when it is handled according to the same protocol used in the evolution experiment. This cohort productivity could be estimated in a *Drosophila* culture, for example, by counting the total number of eggs laid by adults of the experimental population during the reproductive window used to start the next generation and comparing it to the number of eggs laid by a control population used as a proxy for the ancestral state. Counting the number of resulting adult offspring might provide a better estimate of fitness because it would reflect differences in egg viability and larvae survival (e.g., Santos et al., 2023).

A second approach to estimate absolute fitness involves calculating lifetime reproductive success from the underlying life-history components (Lenski & Service, 1982). This can be done, for example, by multiplying a female's age-specific viability by her age-specific fecundity in the relevant environment. As with the cohort-based approach, one would then compare experimental and control populations to assess whether fitness had changed as a consequence of the experimental evolution. This approach leads us to the challenges of assaying life-history traits in the context of experimental evolution.

4.2. Life-History Traits

We will discuss life-history traits in terms of species with females and males with obligate sexual reproduction. However, hermaphroditic and other systems of reproduction can be studied in analogous ways, with the success of female and male reproductive functions combined together to define total fertility. Also, even in asexually reproducing microbes, there are demographic parameters such as growth rates, death rates, and lag times that can be measured and analyzed to see how they contribute to changes in fitness (Atolia et al., 2020; Vasi et al., 1994).

The following life-history traits are commonly measured in sexual species: viability from zygote to reproductive adult, age-specific adult mortality rates, female fecundity, and male virility. We will discuss each of them in turn.

4.2.1. *Juvenile viability*

The survival of juveniles from egg or seed to adulthood is a critically important life-history trait in multicellular species. It is relatively easy to measure, subject to two major provisos.

First, as with all functional traits, it is important to measure viability in the same environment used in the generation-to-generation rearing of the experimental populations of interest. Note, however, that this approach could involve two or more different environments. For example, the rearing environments could be *both* that of the ancestral or control populations *and* that used in one or more of the treatments subjected to a novel selection regime. Such dual assays are often of particular scientific interest because they allow one to test for genotype-by-environment interactions. It is also important to account for possible maternal effects that may confound results. To that end, one can maintain both the experimental and control populations in the same environment for one or two generations prior to the assay, following a "common garden" protocol (Kawecki et al., 2012). Ideally, one could assay both treatment and control populations employing an orthogonal design, with both the ancestral and new environment used for both the preconditioning and assay steps (e.g., Magalhães et al., 2011), although doing so increases the overall size and workload of the assays.

Second, viability is subject to strong density-dependent effects. Thus, standardizing egg or seed numbers relative to resources is critical, even if the normal rearing protocol in an evolution experiment is not standardized, especially if viability is to be compared across multiple populations and generations. Indeed, much can be learned about the nature of adaptation by measuring life-history traits across a range of densities (e.g., Mueller et al., 1993), especially when the selective regimes in an evolution experiment involve manipulating population densities themselves.

4.2.2. *Adult age-specific mortality rates*

In early experiments on life-history evolution, it was common for differences in mortality rates to be based on average longevities (e.g., Luckinbill et al., 1984). At that time, it was thought there were relatively predictable increases in age-specific mortality. However, Carey et al. (1992) and Curtsinger et al. (1992) showed that mortality rates actually plateau late in adulthood, invalidating that assumption. Subsequent work has strengthened the conclusion that there are two different phases to adult mortality (Arnold & Rose, 2023; Mueller et al., 2011; see also Chapter 7). Thus, it is important to study age-specific mortality rates over relatively narrow age windows when studying demographic differences between experimentally evolved populations. Doing so, however, generates two further problems.

First, large cohorts are necessary to measure adult age-specific mortality rates with suitable accuracy. Ideally, one would want to examine thousands of individuals per population. However, few laboratories have the resources necessary to house such large cohorts and quantify death in each age-specific window.

Second, these cohorts must be assayed in the pertinent environment, specifically under those conditions used for and relevant to the evolution experiment itself. For example, Rose et al. (2002) measured age-specific adult mortality under conditions that were not appropriate for most of the populations in that study. When the assay environment was changed to match the conditions of the experimental evolution used for these same populations, radically different results were obtained (Burke et al., 2016), and those new results were more easily understood. Such genotype-by-environment interactions affecting life-history traits have been seen repeatedly in evolution experiments with *Drosophila* (Arnold & Rose,

2023; Leroi, Chen, & Rose, 1994; Leroi, Chippindale, & Rose, 1994; Rose et al., 1996), and they are probably quite general.

An important issue in assays of age-specific mortality is the mating status of adults. It has long been known (e.g., Maynard Smith, 1958) that the mating status of *Drosophila* females has an enormous impact on their longevity, and we expect that this effect is common in many animals. And yet, the typical practice in gerontological research with flies and other laboratory organisms has been to assess longevity using virgin females or males, which can be highly misleading, especially in the context of evolution experiments where reproduction is essential to run the experiment. In general, life-history traits should be assayed under the mating regime that prevailed during the evolution experiment.

4.2.3. *Female fecundity*

As with viability, the population density used when conducting adult life-history assays should be controlled, both for reproducibility and to avoid genotype-by-environment interactions. By controlling density, we mean having the same density across the regimes under study. However, in some experiments, different densities may inevitably arise because of treatment effects. In that case, assaying fecundity across a range of densities would be the best practice. In *Drosophila* experiments, it has been common practice to estimate female fecundity at very low densities, often with only a single pair (male and female) per vial. This approach makes it easy to characterize age-specific changes in fecundity, including estimating the age of first reproduction. Nonetheless, one must be mindful that there may well be large differences in fecundity at such a low density when compared to the high densities that prevailed during the evolution experiment itself (Leroi, Chippindale & Rose, 1994; Teotónio et al., 2002). Similarly, food availability should be carefully controlled. Restricted feeding can have enormous impacts on fecundity and adult mortality (e.g., Chippindale et al., 1993).

As with mortality, age-specific fecundity has distinct adult and late-life phases (Rauser, Abdel-Aal et al., 2005; Rauser, Hong et al., 2005; Rauser et al., 2003). It is not always the case that fecundity simply tails off after a peak in early adulthood. Fortunately, the cohort sizes needed to measure age-specific fecundity are not as large as those required for age-specific mortality. Also, when assessing female reproductive success, it is essential to integrate information on egg viability and female fecundity. This information may be obtained by directly estimating the number of surviving offspring at a given female age or by making independent estimates of the two traits and multiplying them together.

4.2.4. *Male virility*

Assaying male reproductive success is one of the most challenging of all life-history assays. In most species with two sexes, the main factor limiting male reproductive success is the availability of females. The males of most species could fertilize many more females than they actually do under most conditions. This limitation reflects female choice, competition with other males, or both.

This issue raises important questions for designing assays of male reproductive success. Should such assays measure success in the absence of competition with other males and with an abundance of available females? Or should the assays be conducted under conditions in which females have a choice and there are male competitors? Shahrestani et al. (2012) used the former approach and demonstrated the existence of both adult and late-life phases of age-specific male reproductive success in *Drosophila* virility. Thus, male reproductive success, like female fecundity, may exhibit complex age-specific patterns even under simple conditions.

4.3. Functional Traits

As we have just discussed, assays of life-history traits present a wide range of challenges. Many of the same challenges arise with assays of functional traits, especially when they may involve differences among life-cycle phases (e.g., developmental stages) or genotype-by-environment interactions when the experimental organisms alter their own environments (e.g., resource availability). We briefly discuss below functional traits in three broad categories: morphology, physiology, and behavior.

4.3.1. *Morphology*

In unicellular organisms, morphology comes down to cell size and shape. Many bacterial species have relatively simple geometry (e.g., rod-shaped), and in these cases both cell volume and shape can be estimated from microscopic images. Some other bacterial species have more complex cell shapes (e.g., spiral-shaped), while amoebae often have ill-defined shapes due to the absence of a rigid cell wall. Even in these cases, cell volumes can be measured using specialized devices, such as Coulter counters, while computational tools can assist with quantifying more complex aspects of morphology (e.g., surface area).

Cell size and shape often change in microbes under various selection regimes. For example, all 12 *E. coli* populations in the LTEE evolved to have larger cells, and many of them changed shape to become less rod-like compared to their ancestors (Grant et al., 2021). Similarly, experimentally evolved magnetite-resistant *E. coli* became significantly longer and wider than their ancestors (Ewunkem et al., 2021).

To this point, we have discussed microbes as unicellular organisms, but that is not always the case. Some bacteria (e.g., myxobacteria) and eukaryotic microbes (e.g., slime molds) can form aggregations called "fruiting bodies" that promote spore formation and thereby enable some portion of the cells to survive harsh environments. These microbes also aggregate to form predatory "wolfpacks" in the case of myxobacteria and "slugs" that migrate to favorable conditions in the case of slime molds. In a fascinating series of evolution experiments, Ratcliff and colleagues have evolved multicellular yeast starting from a unicellular ancestral strain of *Saccharomyces cerevisiae* (Ratcliff et al., 2012). This multicellularity was selected in a remarkably simple way by gently centrifugating cultures and then propagating the organisms that settled the fastest. In this case, multicellularity evolved not by aggregation, but by successive generations of mother and daughter cells that remained physically attached. The resulting multicellular lines have snowflake-like morphologies; the packing of branches in the snowflakes balances the need for oxygen and resources to diffuse to the cells with the tensile strength required to hold the structure together (Bozdag et al., 2023).

Overall size and shape are both important in multicellular organisms as well. Weight is a common surrogate for overall size. However, it should be emphasized that "wet" and "dry" body weights can be quite different, depending on the degree of desiccation that the organism is subjected to before weighing. In *Drosophila*, another common surrogate for body size is wing length, which is convenient for accurate measurements (Santos et al., 2006).

In terms of shape, multicellular organisms present a staggering diversity of geometries and rigidities. The problems of assaying the shapes of whole multicellular organisms are challenging. A common approach is to focus on a few easily measured morphological traits, such as limb length. Statistical approaches using geometric morphometrics can be used to quantify shape differences. For example, by defining multiple landmarks in the wings of *Drosophila*, one can test for evolved changes in their shape (Santos et al., 2006).

4.3.2. *Physiology and behavior*

Physiological traits have been of great interest in experimental evolution. Indeed, experimental evolution has played an important role in the development of the field of evolutionary physiology (Bennett & Lenski, 1999; Burke & Rose, 2009; Rose et al., 2004). Some of the pioneering physiological and behavioral assays using experimental evolution focused on stress resistance (e.g., Service et al., 1985) and locomotor performance (e.g., Graves et al., 1988), respectively, in *Drosophila* populations that had evolved different patterns of aging. By examining correlations between physiology, behavior, and life history across diverse experimental populations, one can better distinguish between those traits that merely correlate with life history and those that directly enhance Darwinian fitness (e.g., Phelan et al., 2003). Evolution experiments can also be designed that select directly on physiological or behavioral performance (e.g., Rose et al., 1992; Swallow et al., 1998), or indirectly via responses to environmental stresses such as temperature (e.g., Kellermann et al., 2015; Tobler et al., 2015). Such experiments can illuminate relationships among physiological and behavioral traits (e.g., Archer et al., 2007, Swallow et al., 2009), providing an important tool for functional analysis, while also revealing the functional foundations of Darwinian fitness itself.

Research using experimentally evolved populations has drilled down to examine the biochemical underpinnings of adaptation. For example, some metabolites associated with adaptation to stress in *Drosophila* have been identified (e.g., Djawdan et al., 1998; Service, 1987). Going in the other direction, some research has used experimentally evolved lines to study organ evolution, including the *Drosophila* heart (e.g., Shahrestani et al., 2017). It is also now possible to combine the analysis of genome-wide differentiation (see Chapter 5) with physiological and behavioral traits using machine learning (e.g., Kezos et al., 2023). These applications of new technologies to evolution experiments provide interesting opportunities to unravel the genomic foundations of physiological and behavioral functions.

In microbial evolution experiments, physiological research has often focused on possible trade-offs associated with adaptation to growth on different resources or resource concentrations (e.g., Blount et al., 2012, 2020; Dykhuizen & Dean, 2009; Quan et al., 2012; Velicer & Lenski, 1999); growth at different temperatures (e.g., Bennett & Lenski, 1993, 1999; Bennett et al., 1992; Cooper et al., 2001; Tenaillon et al., 2012); and resistance to growth inhibitors and killing agents, including antibiotics, metals, and bacteriophages (e.g., Baym et al., 2016; Bohannan & Lenski, 2000; Card et al., 2021; Chao et al., 1977; Jeje et al., 2023; Lenski, 1988; Meyer et al., 2010; Tajkarimi et al., 2017). (Bacteriophages, or phages for short, are viruses that infect bacteria.) For example, Joseph Graves, Jr., and colleagues found that increased resistance to the metal gallium came at a cost of lower resistance to the antibiotic rifampicin (Graves et al., 2019). However, trade-offs are not universal. For example, Meyer et al. (2010) found that many of the LTEE populations evolved resistance to infection by phage λ (pronounced lambda), despite the evolving *E. coli* not being exposed to λ or any other virus during the experiment. This "trade-up" (as opposed to trade-off) evolved because it was evidently advantageous for the bacteria to reduce expression of an unused transport protein, which happens to be the receptor used by λ to enter cells. Some microbial evolution experiments have been able to probe the mechanistic underpinnings of adaptation more deeply than is typically feasible using multicellular organisms, simply because the effects of genetic variation on physiological functions are often better studied, more direct, and therefore easier to understand in microbes than in multicellular organisms.

Evolution experiments have also been used to investigate how adaptation to specific selection regimes leads to changes in gene expression. In an early transcriptomic study, Cooper et al. (2003) found parallel evolution in the expression profiles of LTEE lines that suggested underlying changes in the *stringent response*, which allows bacteria to cope with alternating periods of growth and starvation. Using the transcriptomic data, the authors sequenced and manipulated candidate genes, which led to the discovery of a beneficial mutation in a key regulatory gene well before whole-genome sequencing was feasible. Since then, new technologies have enabled comprehensive analyses of transcription and translation. Recent work on the LTEE lines identified parallel changes in transcription leading to down-regulation of several functions that are presumably unimportant for the bacteria in the LTEE environment (O-antigen synthesis, siderophore production, flagellar assembly, and secretion systems) and up-regulation of other functions that evidently are important for fitness in that environment, including the synthesis and metabolism of several amino acids, NAD synthesis, and acetate metabolism (Favate et al., 2022). Most functional changes were driven by altered transcription, with changes in translation playing a much smaller role. Changes in gene-expression profiles were also documented in experimental *E. coli* populations that evolved resistance to both copper and iron (Boyd et al., 2022; Thomas et al., 2021).

Mating behaviors are closely related to fitness via their impact on life-history traits, including fecundity in females and virility in males, as discussed earlier in this chapter. In Chapter 7, we will discuss the experimental evolution of wheel-running behavior in mice. However, one might not think about behaviors in the context of microbes. In fact, some microorganisms exhibit fascinating *social* behaviors. For example, as noted in the section on morphology, both myxobacteria and eukaryotic slime molds can form multicellular "fruiting bodies" under certain stressful conditions, which requires cooperation that is mediated by chemical signals. These microbes can also form predatory "wolfpacks" (myxobacteria) and "slugs" that can move to more favorable habitats (slime molds). However, these socially cooperative behaviors are vulnerable to cheaters and misfits (Larsen et al., 2023; Strassmann et al., 2000; Velicer et al., 2000). When experimental populations of *Myxococcus xanthus* evolved for 1,000 generations in an environment where there were no opportunities to engage in the social behaviors, the bacteria lost those abilities (Velicer et al., 1998). Moreover, some of the noncooperators actually gained a fitness advantage when they were mixed with cooperators under conditions where the cooperators produced fruiting bodies (Velicer et al., 2000). In some cases, the disruption caused by cheating was so severe that it caused the extinction of the entire population, including the cheaters and cooperators alike (Fiegna & Velicer, 2003). In one population, a new mutant arose among the cheaters that simultaneously restored cooperation and suppressed the ancestral cheater, preventing extinction (Fiegna et al., 2006). In another experiment, *Myxococcus xanthus* populations were propagated for almost a year on agar plates that selected for improvement in their abilities to search for and consume the patches of *E. coli* that served as their prey (Hillesland et al., 2009).

4.4. Chapter Summary

There are many phenotypes of interest when conducting experimental evolution: Darwinian fitness itself, the life-history traits that together define fitness, and various physiological, morphological, and behavioral traits that contribute to organismal performance. It is important to be aware of the best practices for measuring fitness and life-history traits, as well as the challenges associated with those measurements. The various physiological, morphological, and behavioral traits of interest in a given study may sometimes change repeatably across replicate populations, and other times unpredictably, depending on whether and how they are coupled to fitness.

Chapter 5

Genetic and Genomic Responses to Experimental Evolution

In Chapter 4, we discussed how to measure and evaluate phenotypic responses that occur during evolution experiments. Evolution experiments, especially those of long duration, can produce substantial phenotypic differences between control and treatment groups. Here, we will explain how genomic analysis can reveal the sites of genetic change that are responsible for such phenotypic evolution (Santos et al., 2024).

5.1. Evolve and Resequence

We use the phrase "Evolve and Resequence", or E&R for short, to refer to evolution experiments in which the genetic basis of change is examined at a genome-wide scale (Turner et al., 2011). Resequencing implies that the relevant organism's genome has already been sequenced, establishing a reference for comparison. Thus, sequencing the genomes of the particular ancestor and experimentally evolved control and treatment populations from a species that has previously been sequenced can be considered resequencing.

For most organisms used in evolution experiments, one or more complete genomes of that species are available from the National Center for Biotechnology Information (NCBI). Indeed, the expanding accessibility of both sequencing and bioinformatic methods has enabled in-depth analysis of increasing numbers of diverse species, which has proven valuable not only for experimental evolution but also for comparative studies

57

of natural populations. *De novo* sequencing can also be performed for species or particular strains that are not in the NCBI database, and a variety of methods are available for genome assembly and annotating predicted gene functions (Szalay & Golovchenko, 2015). In fact, with the declining costs and increasing speed and accuracy of genome sequencing, many new evolution experiments use the E&R approach from the start.

5.1.1. *Technical issues*

We will not cover in detail the current state of methods used to sequence genomes, except insofar as the methods may bear on the inferences based on genomic analyses of samples from evolution experiments. There have been tremendous technical advances in recent years, and further changes will likely continue into the future. Prior to the first decade of the 21st century, it was difficult to study the genomic basis of adaptation owing to limitations of the available DNA sequencing technology. The Sanger sequencing method that prevailed in the late 20th century was laborious and costly. By contrast, "next-generation sequencing" (NGS) enabled massively parallel and high-throughput analysis of multiple samples at a greatly reduced cost (Mardis, 2011). The success of NGS was achieved by miniaturizing sequencing reactions, employing microfluidics, and developing highly sensitive detection systems (Kulski, 2016).

Among these NGS methods, Illumina platforms became an industry standard in the first decade of the 21st century. A drawback to the Illumina method, however, is that it works by cutting the DNA into small fragments of ~300 bp (at present) and then reading the fragments separately. These reads are then aligned to a reference genome using computational methods in order to identify differences and thereby produce the new sequence. This approach works well for genomes that do not contain many long repetitive stretches, such as those of viruses and bacteria. However, it cannot properly arrange the order of repetitive portions of the genome, which are common in most eukaryotic genomes. One way to circumvent this problem is by using paired-end sequencing reads, which provide high-quality alignments across DNA regions with repetitive sequences. This approach produces long overlapping reads (called "contigs") that allow one to bridge gaps in the genome sequence; it is especially important for *de novo* sequencing or to identify point mutations and rearrangements that affect repetitive regions of the genome.

The cutting of genomic DNA into small fragments means that each stretch of DNA will generally be sequenced many times. The number of times that a genomic position is sequenced, on average, is called the "depth of coverage." The deeper that coverage, the higher the confidence that a particular variant in the genome is real and not the result of sequencing error. At present, a 20-fold coverage is usually viewed as acceptable, while 50-fold is considered the gold standard. It is also important to review reports from the quality and read-filtering algorithms, which indicate the estimated error rates. The lower the quality score, the higher the error rate. With Illumina sequencing, a quality score of 30 indicates a probability of an incorrect base call of about one in a thousand. While this accuracy is impressive, these errors underscore the importance of depth of coverage given the millions or even billions of sites in most genomes.

Newer methods that provide much longer sequencing reads have also been developed (van Dijk et al., 2018). For example, the PacBio and Oxford Nanopore technologies generate read lengths of more than 10 kilobases (kb). However, these longer reads generally have much higher error rates, though progress is being made in reducing these rates (Tyler et al., 2018). Studies are increasingly using a hybrid approach that combines the best of both worlds, taking advantage of the long-read technologies to span repetitive sequences and thereby assemble complete genomes, on the one hand, and exploiting the high coverage provided by Illumina to discover mutations and reduce the overall error rate, on the other hand (Wick et al., 2023).

5.1.2. *Sequencing individuals and populations*

In analyzing an evolution experiment, one can sequence either individual organisms or samples that contain many individuals from a population. Sequencing individuals provides the advantage of being able to directly determine the genetic linkage between evolved variants. Doing so might also allow one to associate specific phenotypes with potentially causative genetic variants. A drawback, however, is the cost of sequencing many individuals, especially given the sample sizes required and the population-level replication that should be present in any well-designed evolution experiment.

With microbes, it is feasible to sequence a large bulk sample of an experimental population containing millions of cells (Barrick & Lenski, 2009; Good et al., 2017), either alone or in combination with sequencing individual clones from the same sample (Barrick et al., 2009; Tenaillon et al., 2016). By sequencing the population sample, one does not get an exact genome for any individual organism. Instead, one gains information about two sorts of new mutations that may have arisen in the experiment. First, some mutations might have been fixed in the evolving population, reaching a frequency of 100% in the sample. Second, other mutations might be present at an intermediate frequency, indicating a polymorphism in the population.

Some polymorphisms are maintained by *negative frequency-dependent selection*, which favors rare variants, or, in the case of diploid organisms, by a *heterozygotic advantage* in which heterozygotes have higher fitness than homozygotes. Other polymorphisms might be selectively neutral or weakly deleterious "passenger" mutations that can increase in frequency by *hitchhiking* along with genetically linked beneficial "driver" mutations. Still other polymorphisms might be transient, such as when a beneficial mutation is on its way to fixation but has not yet reached 100% frequency.

For multicellular organisms, sequencing the DNA extracted from a pool of many individuals ("Pool-Seq") from an experimental population (Schlötterer et al., 2015) offers a similar approach to bulk sequencing of a microbial population. Pool-Seq can be cost-effective for large populations, and it allows one to estimate genome-wide allele frequencies and even the frequencies of haplotypes within the population (Rode et al., 2018).

The programs used to analyze genomic data generally start with quality control and read filtering, followed by alignment to the reference genome, and then "calling" (identifying) genetic variants. A widely used pipeline for bacterial resequencing projects is called *breseq* (Deatherage & Barrick, 2014). It is an open-source, command-line tool implemented in C++ and R that runs on a variety of UNIX platforms. The *breseq* pipeline calls single nucleotide polymorphisms (SNPs), small "indels" (insertions and deletions of one or a few bases), large deletions, and new junctions produced by large insertions, duplications, and inversions. The output is provided in annotated HTML format. Several variant-calling algorithms have been developed for Pool-Seq data obtained from diploid organisms (reviewed by Guirao-Rico & González, 2021). For example, PoPoolation is written in Perl and R, and it builds on commonly used data formats (Kofler, Orozco-terWengel et al., 2011; Kofler et al., 2011). This pipeline uses BWA or Bowtie for alignment, filters ambiguously mapped reads using SAMtools, and then calls variants that differ from the reference genome. A related program, PoPoolationTE, can be used for transposable genetic elements (Kofler et al., 2016). Vlachos et al. (2019) tested and evaluated a number of alternative software tools for analyzing Pool-Seq data.

5.2. Identifying Beneficial Mutations

After analyzing the genome sequences from an evolution experiment, one typically has for each population a table of new mutations, in the case of experiments with asexual microbes, or of alleles that underwent significant increases in frequency, for multicellular sexual organisms. What then?

One might be tempted to assume that any new mutation that fixed in an asexual population, or any existing allele that increased substantially in frequency in a population of sexual organisms, must have been beneficial under the conditions of the experiment. However, it is critically important to understand that new or existing alleles can increase in frequency and even reach fixation without being selectively advantageous. In particular, neutral and even slightly deleterious alleles may increase in frequency by random drift and especially by hitchhiking with beneficial mutations (see Figure 5.1). Such hitchhiking can occur in both asexual and sexual systems.

In asexual organisms, the entire genome functions as a single linkage unit. Thus, when a new beneficial mutation occurs in a particular clone, any alleles already present in that clone will accrue the same net fitness advantage as the new beneficial mutation. For example, if a mutation that confers a 10% fitness advantage occurs in a clone that harbors a weakly deleterious mutation with a 1% disadvantage, both the beneficial and deleterious mutations will experience a 9% advantage relative to some other clone that lacks both mutations. The situation is a bit different if the order

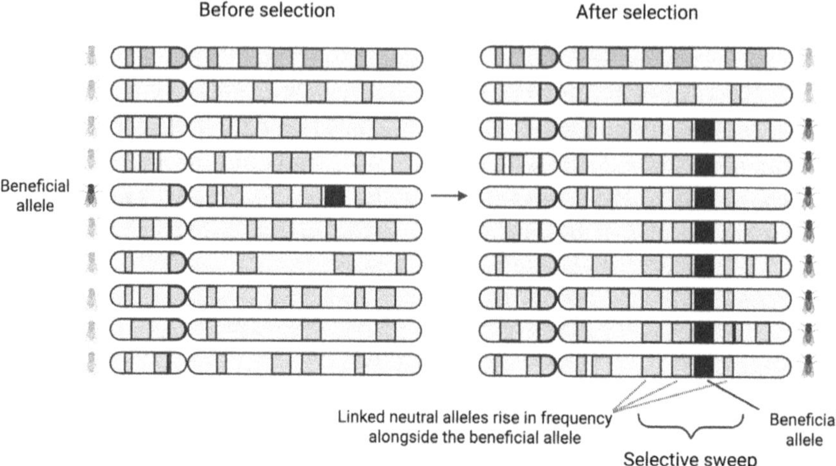

Figure 5.1:   Selective sweeps reduce genetic variation in the vicinity of beneficial mutations. Genetic variation in the vicinity of beneficial mutations is reduced because new or existing alleles can increase in frequency and even reach fixation without being selectively advantageous. In particular, neutral and even slightly deleterious alleles may increase in frequency by random drift and especially by hitchhiking with beneficial mutations.

of mutations is reversed, such that the deleterious mutation occurs a few generations *after* the beneficial mutation. In that case, both the clones with and without the deleterious mutation will increase for some period relative to a clone without either mutation, but eventually the clone with only the beneficial mutation should slowly out-compete the one that also carries the deleterious mutation.

In sexual organisms, the potential for hitchhiking is somewhat reduced by genetic recombination, which tends to erode genetic linkage over time. However, most sexual multicellular organisms have larger genomes and longer generations than asexual microbes. In these species, evolution experiments typically begin with abundant standing variation. As a consequence, it is often more challenging to distinguish between beneficial driver alleles and neutral or deleterious hitchhikers in experiments with sexual populations, despite the effects of recombination.

In asexual microbes, there are two main approaches used to identify beneficial mutations. The first approach is based on finding *parallel* genetic changes, which are similar or even identical mutations that have arisen in multiple experimental populations (Tenaillon et al., 2016; Woods et al., 2006). If each asexual population was founded from a single colony using the outgrowth of a single cell (as discussed in Chapter 3), and assuming that cross-contamination did not occur, then it would be highly unlikely to find the exact same mutation rising to fixation repeatedly (i.e., in parallel) *unless* the mutation conferred a large fitness advantage. Such precisely identical mutations tend to be rare in microbial evolution experiments, but they have been observed in a few studies (e.g., Meyer et al., 2012). It is more common to see repeated mutations that are similar—in the same genes, operons, or pathways—but not identical at the nucleotide level. Establishing whether parallel changes are unexpectedly common (implying the mutations are beneficial) or merely coincidental requires rigorous statistical tests. For example, in the Long-Term Evolution Experiment (LTEE) with *Escherichia coli*, Tenaillon et al. (2016) found that half of all the observed *nonsynonymous* mutations (excluding lineages that had evolved hypermutability) were found in a set of genes that comprised only 2% of the protein-coding genome. (Nonsynonymous mutations are point mutations that change one of the amino acids in a protein.) A randomization test indicated that this concentration was highly significant, thus supporting the inference that these mutations were indeed beneficial.

The second approach to identifying beneficial mutations with asexual microbes is to use genetic methods to make isogenic constructs—that is, two strains that are *identical except* for a mutation of interest—and then compete them under the conditions of the evolution experiment. For example, Crozat et al. (2005) discovered that most LTEE lines had evolved changes in the supercoiling of their chromosomes, and they found mutations in two genes, *topA* and *fis*, known to affect that trait. They then moved each mutation into the ancestral genetic background and performed competitions between strains with and without the evolved mutations. They found that both mutations indeed conferred a significant fitness advantage. This approach has been used with a number of other mutations from the LTEE, and most of those tested were also shown to be beneficial (Barrick et al., 2009; Khan et al., 2011). A potential complication may arise, however, if there are strong *epistatic interactions* between the mutation of interest and other mutations that arose during the experiment. (Recall that epistatic interactions indicate that the combined effects of two or more mutations cannot be inferred from their separate effects.) For example, a mutation may have been beneficial in the genetic background in which it arose, but not when it is moved into the ancestral strain. That possibility can be tested by making and competing a pair of isogenic strains in which the ancestral allele replaces the mutation in an evolved strain.

The same two lines of evidence—testing for parallel changes and making isogenic constructs—can also be used with sexually reproducing organisms, though with some additional considerations. On the one hand, the presence of abundant genetic variation in the ancestral population can lead to widespread hitchhiking, thus making it difficult to pinpoint exactly which genetic differences are responsible for any observed parallelism. That linked variation can also make it difficult to move only a single mutation in order to make isogenic constructs and test their fitness effects. On the other hand, distinct but overlapping blocks of linked alleles may have increased in frequency or fixed in the different replicate populations, which may facilitate identifying the causative allele statistically. Also, sexual reproduction can be useful when making isogenic (or near-isogenic) constructs because repeated backcrosses can help zero in on the causal alleles, at least in the absence of severe epistasis.

5.3. Best Practices in E&R Research

5.3.1. *Replication*

In E&R experiments, as in any evolution experiment, having more replicate populations will generally strengthen one's statistical inferences. However, it is more costly to maintain more evolving populations, and while sequencing costs have declined greatly in recent years, those costs should be kept in mind when designing E&R studies. Having said that, not every population and sample timepoint necessarily needs to be sequenced.

5.3.2. *Controls*

E&R studies require sequencing both evolved treatment populations and the ancestors or other control populations. If one uses an ancestral population obtained from another source (e.g., a stock center or commercial enterprise), then it is important to verify that the organisms you received are actually those that you requested. This verification is usually straightforward with larger organisms, but it can be a bit more difficult with microbes. A simple technique for microbes is to employ serial dilution and plating, such that each colony is derived from a single cell. In some cases, colony appearance may suggest that a sample is contaminated with other microbes. In any case, one can isolate and sequence one or more colonies to ensure that one has the desired organism before starting an evolution experiment.

We are aware of some instances in which laboratories received either the wrong microbial species or a mixture of the intended species along with a contaminant. Even when an uncontaminated culture of the proper species is obtained, it is worthwhile to sequence its genome because mutations may have occurred since the strain was sequenced and the information deposited in a database. For example, different labs and culture collections possess genetically distinct sub-strains of the canonical *E. coli* K–12 (Graves et al., 2015; Naas et al., 1995) and *E. coli* B strains (Jeong et al., 2009) that diverged over time in the course of their handling and propagation.

During an E&R experiment, control populations (derived from the founding ancestor) are likely to adapt to the control environment, including the culture medium (e.g., the banana-molasses food often used for fruit flies). Genetic variants that sweep to high frequency in *both* control and treatment populations indicate adaptation to shared aspects of the control and treatment environments, which may include food, temperature, or any other feature in common to those environments. Conversely, variants that sweep only in the treatment populations (and not in the controls) provide evidence of adaptation to the particular features of the treatment conditions.

5.3.3. *Genetic variation*

In Chapter 3, we discussed the possibility that parasexual processes may occur in experiments with otherwise asexual microbes, which could allow mutations that arose in different lineages to recombine. In the ancestral *E. coli* used to start the LTEE, these processes do not occur. Thus, the results of E&R reflect the effects of selection and drift acting on new

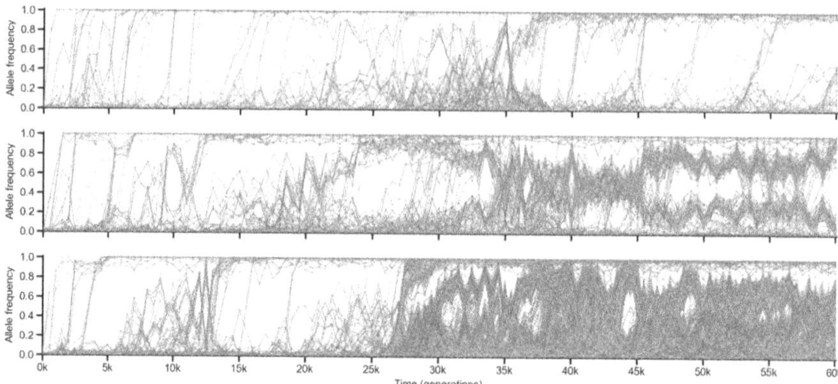

Figure 5.2: Trajectories of allele frequencies in three representative LTEE populations across 60,000 generations. Whole-population samples were sequenced at 500-generation intervals. Each panel shows the trajectories of all mutations that reached sufficiently high frequency to be detected (above ~5%). Some mutations eventually fixed (i.e., reached 100% frequency), but others did not. Top panel: This population underwent a series of selective sweeps culminating in the fixation of mutations. Middle panel: This population started in a similar manner before diverging into two subpopulations, or lineages, that then coexisted for tens of thousands of generations. Bottom panel: This population evolved hypermutability after about 25,000 generations, resulting in huge numbers of mutations that are difficult to visualize (adapted from Figure 1 in Good et al. [2017]).

mutations (Lenski, 2023). Early on, the LTEE populations exhibited a pattern of "hard" selective sweeps, in which lineages that harbored beneficial mutations replaced their progenitors and those mutations went to fixation (see Figure 5.2, top panel). In some cases, the metagenomic data also revealed the effects of *clonal interference*, whereby lineages with the most highly beneficial mutations outcompete not only the progenitors but also lineages with somewhat less beneficial mutations that increase in frequency before ultimately going extinct. In later generations, the metagenomic data became even more complex in many populations owing to two additional evolutionary phenomena. First, some LTEE populations evolved into two lineages with distinct *ecotypes* that coexisted owing to *negative frequency-dependent selection* (Figure 5.2, middle panel). That is, each ecotype grows faster than the alternative type when it is rare, but not when it is common, with the result that the two ecotypes can coexist. Over time, beneficial mutations continued to sweep through each lineage, perturbing the frequencies of the two ecotypes and, in some but not all cases, driving one or the other type to extinction (Maddamsetti et al., 2015; Rozen & Lenski, 2000). Second, some populations evolved *hypermutability* (elevated mutation rates) caused by changes in DNA repair and metabolism. These populations accumulated much greater genetic diversity, which tended to obscure later selective sweeps (Figure 5.2, bottom panel).

In another experiment, bacteria derived from the LTEE after 10,000 generations were propagated for an additional 1,000 generations in the same environment except the treatment populations (but not the controls) experienced periodic conjugation with donors derived from a different *E. coli* strain (Souza et al., 1997). These donors were *auxotrophic mutants* that could not grow in the minimal medium because they could not produce a required amino acid, but they could nonetheless transfer genetic material to the LTEE-derived cells. The goal of the experiment was to test the hypothesis that this parasexual process would increase the rate of adaptive evolution by providing an additional source of genetic variation. However, the results did not support that hypothesis; in fact, there was a slight, but not significant, tendency for the treatment populations to be less fit than the controls (Souza et al., 1997). Subsequent genome sequencing showed that the recombination rate had been so high, in fact, that the introgression of donor alleles overwrote some beneficial mutations that had arisen in the recipient populations during the previous 10,000 generations (Maddamsetti & Lenski, 2018).

We have discussed how, in sexual organisms, Mendelian processes generate abundant genetic variation by segregation and recombination alone, so long as these populations are outbred and not fully homozygous. These systems also generate variation via *de novo* mutations, but evolution experiments using *Drosophila*, *Mus*, and other animals are usually driven mostly by the abundant genetic variation in the founding population that can quickly respond to selection (Burke, 2012). Indeed, E&R studies have shown that the initial responses to selection are often rapid, provided that selection is sufficiently strong (Burke et al., 2010; Graves et al., 2017).

5.3.4. *Evolutionary time*

E&R experiments are more powerful in showing adaptive processes when they run for more generations. In this respect, the LTEE is the gold standard for asexual microbes—it has been running for over 35 years and more than 80,000 bacterial generations. Unicellular eukaryotes such as yeast also have great potential for long-term E&R experiments (Burke, 2023; Johnson et al., 2021; Lang et al., 2013; Phillips, Briar et al., 2022). The longest sustained E&R study using *Drosophila* was performed in the Rose laboratory, with some populations having evolved for about 1,000 generations (Graves et al., 2017).

5.4. Genomic Insights from E&R Studies

Over the past decade, E&R experiments have powerfully demonstrated Dobzhansky's (1973) dictum that "nothing in biology makes sense except in the light of evolution." These experiments have been used to address problems in both asexual and sexual organisms, including adaptation to novel environments, changes in physiological tolerance, resistance to pathogens and toxins, mating behavior, reproductive isolation, life history evolution, aging, and the coupling of phenotypic and genomic evolution (Barrick et al., 2009; Blount et al., 2020; Boyd et al., 2022; Burke, 2023; Deatherage & Barrick, 2021; Good et al., 2017; Graves et al., 2017; Kawecki et al., 2012; Schlötterer et al., 2015; Tenaillon et al., 2016; Waters et al., 2015). By taking samples from E&R experiments at multiple time points, one can also examine the trajectories of particular alleles (Good et al., 2017; Orozco-Terwengel et al., 2012; Phillips, Arnold et al., 2022; Seabra et al., 2018; recall examples in Figure 5.2).

Not surprisingly, analyses of E&R experiments reveal important differences between the dynamics of asexual and sexual organisms. In the LTEE, there was a gradually declining rate of fitness gain as the populations adapted to the low-glucose environment. Meanwhile, the average genetic distance relative to the ancestor (i.e., average number of mutations in an evolving population) continued to increase (Barrick et al., 2009; Good et al., 2017; Tenaillon et al., 2016). In the populations that evolved hypermutability, the pace of genomic change accelerated sharply. In the populations that retained the low ancestral mutation rate, the pace of genomic evolution declined over time, but more slowly than the decline in the rate of fitness improvement. The more gradual decline in the rate of genomic change reflects, in part, the fact that neutral mutations continue to accumulate even as adaptation slows. By contrast, an E&R experiment with outcrossing yeast under selection for greater ethanol tolerance did not find the steady fixation of genetic variants (Phillips, Briar et al., 2022). Instead, that study found distinct adaptive responses to different selection intensities, including some variants that were associated with general adaptation to the lab environment and others with specific adaptation to different levels of ethanol.

In an evolution experiment using *Drosophila melanogaster*, Orozco-Terwengel et al. (2012) identified two contrasting types of dynamics in SNPs thought to be responsible for adaptation; some quickly reached a plateau, while others showed a slow and steady increase. Using *D. subobscura*, Seabra et al. (2018) also saw diverse genome dynamics, including even reversals in the direction of frequency change in some SNPs. The different patterns in these studies might reflect different experimental regimes, different species or source populations, or even different statistical methodologies. This last possibility suggests that more attention should be paid to standardizing the protocols used for data analysis in order to ensure the reproducibility of results from E&R studies.

E&R studies in *Drosophila* have tended to find large numbers of SNPs associated with substantial increases in the frequency of selected alleles (Burke et al., 2010; Graves et al., 2017; Nuzhdin & Turner, 2013). The most likely explanation for this phenomenon is *hitchhiking*, in which driver alleles (i.e., those that improve fitness) bring along linked variants that are selectively neutral or nearly so. Hitchhiking is especially prevalent within chromosomal inversions, where recombination between heterozygotes (with and without the inversion) is suppressed. Mueller et al. (2018) developed an approach for distinguishing the causal *driver* alleles from noncausal *passenger* variants in *Drosophila* E&R experiments that uses a type of machine learning called the "fused lasso additive method" (FLAM). They deployed FLAM to parse SNPs from Pool-Seq databases from experimentally evolved populations to identify their associations with life-history traits, including development time, early and late fecundity, and adult mortality. Using two databases with 20 and 30 populations, they found that 53 and 31, respectively, of 194 significantly differentiated SNPs were associated with six phenotypes that had diverged significantly (Mueller et al., 2018). In a related study, Kezos et al. (2023) used FLAM to identify likely causal SNPs from 60 experimental populations for differentiated physiological traits, including starvation survival, desiccation survival, lipid content, and body mass. They found 142 causal SNPs associated with at least one of the eight traits examined. Their results indicated that each trait was affected by multiple SNPs and, moreover, that some SNPs likely had *pleiotropic effects*, where the same genetic variant affects two or more traits.

Many E&R studies have reported parallel genomic evolution. Such parallelism can occur at the level of the nucleotide sequence, affected genes, or integrated functions. Parallelism at the sequence level tends to be rare because a single nucleotide is a small mutational target and because different mutations in the same gene often produce similar benefits (Tenaillon et al., 2016). Thus, parallelism is typically found at the level of genes, as observed in the LTEE (Maddamsetti et al., 2017; Tenaillon et al., 2016; Woods et al., 2006). The genetic underpinnings of adaptation to high temperature were also examined in *E. coli* populations that were spun off from the LTEE (Tenaillon et al., 2012). That study found over 1,000 mutations in 115 replicate populations that evolved for 2000 generations at 42.2°C. The replicates did not converge on the same mutations; instead, parallelism occurred at the level of the genes and functional complexes involved in the evolution of thermal tolerance. The authors also found that the replicate populations evolved along two distinct adaptive trajectories, each characterized by mutations in different genes (Tenaillon et al., 2012). By contrast, Tajkarimi et al. (2017) found extreme parallelism even at the nucleotide level during the adaptation of *E. coli* to ionic silver. In this E&R experiment, 13 of the 19 populations that were sequenced had mutations in the *cusS* gene, which encodes a protein involved in sensing metals in the periplasm. In fact, many had the exact same mutation that replaced an arginine with a lysine at the 15th amino acid in the protein (R15L). Simulations and experiments indicated that the R15L variant was superior at binding ionic silver compared to the ancestor and to other observed mutations. As discussed in Chapter 3, however, it is unclear whether the identical R15L alleles resulted from independent mutational events or, alternatively, derive from a single mutation that spread between replicate populations by accidental cross-contamination.

E&R experiments with sexually reproducing organisms also often exhibit parallel evolution. With sexual organisms, however, most parallel genetic changes result from increases in the frequency of alleles that were already present in the ancestral population. Stern (2013) calls this *collateral* evolution, in contrast to parallel and convergent evolution that reflect truly independent mutational events. In any case, Joseph Graves, Jr., and colleagues performed "replay" experiments with fruit flies modeled on Gould's fanciful idea of "replaying the tape of life" (Blount et al., 2018; Gould, 1989; Graves et al., 2017). To that end, they used as ancestors the well-studied *D. melanogaster* stocks that had evolved altered life-history traits (Rose et al., 2004), and they subjected new experimental populations to repeated shifts in the ages during which reproduction was allowed. In addition to parallel changes in life-history traits, Graves et al. (2017) documented parallel genomic changes involving SNPs, indels, and transposable elements.

E&R experiments have also documented the importance of *epistasis* and, relatedly, *historical contingency*, in which earlier events affect the future path of evolutionary change. For example, Good et al. (2017) showed that some genes tended to accumulate beneficial mutations early in the LTEE but not later on (e.g., *hslU*, which encodes an ATP-dependent protease); others showed the opposite tendency (e.g., *atoS*, which encodes part of a two-component regulatory system that promotes short-chain fatty-acid catabolism). In another E&R experiment, Card et al. (2019, 2021) used clones from the LTEE to start new populations that were challenged with several antibiotics (ampicillin, ceftriaxone, ciprofloxacin, and tetracycline). They found that clones from different LTEE lines—each with a unique genetic background—were predisposed to evolve resistance via mutations in different genetic targets, indicating epistatic interactions between the backgrounds and the target genes.

Similar results have been shown for multicellular sexual organisms. An evolution experiment with a multi-parental stock of hermaphroditic *Caenorhabditis elegans* found that *sign epistasis*—where a mutation flips from beneficial to deleterious, depending on the genetic background in which it occurs—was an important determinant of both fertility and body size at the time of reproduction (Noble et al., 2017). The prevalence of epistasis is supported by long-term domestication studies as well as E&R experiments (Csilléry et al., 2018).

5.5. Transcriptomics

Pool-seq methods can also be used to extract RNA for studies of gene expression (Todd et al., 2016). Whole-transcriptome sequencing ("RNA-seq") supplanted microarray-based studies in the last decade. RNA-seq performs well with regard to its sensitivity to rare transcripts, splice variants, and microRNAs. Owing to typical biological variation in gene expression, RNA-seq studies are even more susceptible to the issues impacting pool-seq. Thus, larger sample sizes and replication are required to achieve the same statistical power as a comparable DNA sequencing study.

Best practices for an RNA-seq study include sequencing many replicate populations, sequencing each sample to at least 10× coverage, and sequencing at least three technical replicates per population. It must also be emphasized that RNA-seq examines phenotypic variation, and therefore great care should be taken in the protocols for sample collection, the environment in which organisms develop, their age and mating status, and whether one collects sample material from a whole organism or only from specific tissues. If one is planning a large RNA-seq study, it is advisable to run a smaller pilot study to work through any problems that might arise at a larger scale. These problems may involve such issues as the age and storage of samples (RNA degrades rapidly if not stored at low temperatures) as well as the specific procedures used in sequencing the RNA.

Bioinformatic analyses of RNA-seq data begin with algorithms that evaluate read quality and filter the data accordingly, followed by mapping the reads to the reference genome or transcriptome, and ending with statistical tests for differential expression (Todd et al., 2016). These tests generally consist of normalizing RNA counts to the average expression of a set of core "housekeeping" genes. The normalized counts are then typically converted to logarithmic ratios ("fold changes"), usually followed by pairwise difference tests that can be visualized as heat maps showing differential expression, often with hierarchical clustering to indicate similar patterns of differential gene expression (Barter et al., 2019; Boyd et al., 2022; Thomas et al., 2021). Thus, some genes will be upregulated or downregulated relative to the housekeeping genes in experimentally evolved control or treatment populations in the test environment. Of course, care must be taken when interpreting RNA-seq data because differential expression may result not only from mutations in or adjacent to the expressed genes but also from differences in other genes that interact with the differentially expressed genes.

E&R experiments have also shown that changes in gene-expression profiles are often associated with new or increased frequency of beneficial alleles in both asexual and sexual organisms (Barter et al., 2019; Carolus et al., 2021; Cooper et al., 2003; Favate et al., 2022; Thomas et al., 2021). As noted previously, Cooper et al. (2003) showed that a mutation that affected the stringent response provided a substantial fitness benefit in the LTEE while also altering the expression of many other genes. Later work on the LTEE lines using RNA-seq found many more changes in the transcriptome, including greater total levels of mRNA compared to the ancestor (consistent with the observed increases in cell volume) and parallel changes in the expression of specific genes. As discussed in Chapter 4, the changes include upregulation of some functions that are likely important for fitness in the glucose-limited LTEE medium and downregulation of other functions that are presumably unimportant in the flask environment. Thomas et al. (2021) examined changes in gene expression in an experiment with *E. coli* that favored increased resistance to excess iron in the medium. They found that the resistant populations had downregulated iron-transport genes while upregulating genes involved in protection against oxidative stress. By contrast, the control populations did not downregulate iron uptake, and they showed only slightly increased expression of the genes involved with protection against oxidative stress (Thomas et al., 2021). An experiment with the yeast *Candida auris* found altered patterns of gene expression against various antifungal compounds (Carolus et al., 2021). For example, resistance to azole was associated with a mutation in a transcription factor that led to overexpression of an efflux pump. Barter et al. (2019) found that hundreds of genes differed in their expression levels in *Drosophila* populations selected for early versus postponed aging. These studies nicely illustrate how experimental evolution can help to identify the genomic basis of adaptation and couple genetic to phenotypic changes.

5.6. Chapter Summary

This chapter examines the value of combining experimental evolution with new genome-sequencing technologies, often referred to as "evolve and resequence" studies, or E&R for short. We present best practices for E&R study design, including the importance of replication, controls, genetic variation, and evolutionary time. It is important to keep in mind that the processes that can cause genetic changes in evolution experiments include not only natural selection, mutation, and recombination but also random genetic drift and hitchhiking caused by genetic linkage. We discuss ways to distinguish between beneficial "driver" mutations and "passenger" alleles that spread by hitchhiking, and we discuss how asexual versus sexual reproduction can affect those inferences. We conclude this chapter by discussing exemplary studies that have used E&R to illuminate the genomic basis of adaptation, test specific evolutionary hypotheses, and connect genetic and phenotypic changes.

Chapter 6

Experimental Evolution in Asexual Populations*

This chapter discusses evolution experiments using asexually reproducing microbes, with a particular focus on the iconic *Escherichia coli* Long-Term Evolution Experiment (LTEE) that was started by Richard Lenski in 1988. It continues to this day, having recently surpassed 80,000 generations under its new director, Jeffery Barrick.

The LTEE addresses three overarching sets of questions. The first set of questions concerns the repeatability of evolution, an issue of longstanding interest to biologists. There is a fundamental tension between the process of natural selection, which systematically favors particular traits depending on their fit to the environment, and the random processes of mutation and genetic drift. The LTEE can examine the repeatability of evolution because there are 12 replicate populations, all founded from the same ancestral strain and then propagated under the same environmental conditions. One possibility is that adaptive evolution might be repeatable, at least in broad outline, owing to the power of natural selection to find similar solutions to the same challenges and opportunities. Alternatively, evolution might be highly unpredictable and idiosyncratic, with each population finding different solutions as a consequence of the random nature of mutations and the contingent history of each population as it accumulates a unique set of mutations. Importantly, the LTEE has allowed the examination of the extent of repeatability for many different traits as

---

*Portions of this chapter have been adapted from Lenski (2023).

well as the factors that influence the extent of parallelism or divergence for a given trait. That is, some phenotypic and genomic changes might be repeatable and others haphazard, depending on the relative importance of natural selection and random processes for particular traits and genes.

The second set of questions addressed by the LTEE concerns the tempo and mode of evolutionary change. Is evolution invariably gradual, or are there episodes of faster and slower change even when the environment is constant? One possibility is that phenotypic evolution is generally slow and gradual because it involves many mutations, with each one having a small effect. Alternatively, phenotypic evolution might exhibit punctuated dynamics that result from the rapid substitution of mutations that have large effects. The LTEE also examines how the dynamics may depend on the particular phenotypic and genetic traits that are measured. Another question is whether fitness improvements can continue indefinitely, or whether some limit on fitness must eventually be reached.

The third set of questions that the LTEE seeks to address concerns the integration of phenotypic and genetic evolution. Are the dynamics of phenotypic and genomic change tightly coupled and thus concordant in time? Or are there periods when genomic evolution continues apace, even while phenotypic evolution is at a standstill? If phenotypic traits evolve in parallel across replicate populations, does that imply parallelism at the level of mutations, genes, or pathways? Are the bacteria becoming ecological specialists as they adapt to the constant environment of the LTEE? And if specialization does evolve, is it driven primarily by pleiotropic trade-offs or by the accumulation of neutral mutations?

6.1. Experimental Design of the LTEE

In Chapter 2, we explained the four essential features of any evolution experiment: replication, controls, genetic variation, and ample evolutionary time. In Chapter 3, we expanded that discussion to present considerations of experimental design for each of two broad categories of organisms that are widely used in evolution experiments: asexually reproducing microbes and sexually reproducing multicellular organisms. We will now discuss in greater depth some aspects of experimental design that are of particular importance for the LTEE.

6.1.1. *Frozen fossil record*

One of the most important features of many microorganisms for evolution experiments is that they can be frozen and then later revived. The ancestral strain and samples from the evolving populations have been periodically frozen throughout the LTEE. This capacity to freeze and revive living cells allows "time travel" in an evolutionary sense. For example, one can revive and compete organisms that lived at different times in order to measure their relative fitness.

Moreover, the capacity to freeze and revive living organisms has enabled Lenski and colleagues to perform analyses of phenotypic and genetic changes using technologies that did not exist when the LTEE started. For example, no whole genomes had been sequenced for any organism when the LTEE began in 1988, and it was difficult and costly even to sequence individual genes. Genome sequencing and other technologies have greatly expanded the capacity to pursue the integration of phenotypic and genetic evolution, as discussed in Chapter 5.

6.1.2. *Population density*

The LTEE uses a minimal medium, called DM25, that contains glucose as the source of carbon and energy for the cells along with various salts. (In addition to glucose, DM25 also contains citrate, which is another potential source of carbon and energy. However, *E. coli* is generally unable to grow on citrate in the presence of oxygen, and it is included in the medium as a chelating agent. We will return to citrate later in this chapter.) The glucose concentration is only 25 mg/L, which is low by microbiological standards, and the bacteria reach a stationary phase density of only ~5 × 10^7 cells/mL. By comparison, bacteria grown in a nutrient-rich broth can reach densities ~100-fold higher. The resulting LTEE cultures are only slightly turbid to the eye, and there is no perceptible turbidity immediately after the daily dilution in fresh medium. With a culture volume of 10 mL, each population contains ~5 × 10^8 cells after the glucose has been completely consumed. This relatively small population size might be expected to slow the rate of adaptive evolution in comparison to a larger population, at least if one assumes that beneficial mutations are rare.

However, beneficial mutations are not rare in the LTEE, as Lenski knew from shorter experiments that had already been performed with the same strain and under similar conditions (Bouma & Lenski, 1988; Lenski, 1988). Indeed, based on a mathematical analysis of the dynamics of fitness trajectories observed in the LTEE populations (Wiser et al., 2013), it has been estimated that the rate of beneficial mutations is about 1.7×10^{-6} per genome. That rate implies that roughly 850 beneficial mutations occur in each LTEE population every day, despite the low glucose concentration. About 99% of those mutations are lost during the 100-fold dilutions but even so, that leaves 5–10 new beneficial mutations that survive each day. Of those survivors, most are eventually eliminated by *clonal interference*, whereby the beneficial mutations with the highest fitness outcompete somewhat less beneficial mutations (de Visser et al., 1999; Gerrish & Lenski, 1998). In any case, there are plenty of beneficial mutations to fuel adaptation in the LTEE, despite the meager glucose concentration and small culture volume (Deatherage & Barrick, 2021; Lenski et al., 1991; Lenski & Travisano, 1994; Wiser et al., 2013).

The low population density in the LTEE was deliberately chosen in order to reduce the potential for metabolic byproducts to accumulate in the cultures. As bacteria grow, they may secrete byproducts that provide secondary resources or, in the case of toxic byproducts, inhibit growth. The presence of byproducts can lead to *frequency-dependent selection*, which can complicate the measurement of relative fitness. By having a low population density, the concentration of any byproducts should also be low, thereby reducing their importance and minimizing the likelihood that complex ecological interactions would evolve. Moreover, the fact that the populations are diluted 100-fold each day means the concentration of any byproduct is even lower while the cells are growing exponentially.

Nonetheless, frequency-dependent selection did emerge in many of the LTEE populations (recall Figure 5.2, middle panel), at least over portions of their evolutionary history (Good et al., 2017; Rozen & Lenski, 2000). However, the frequency-dependent effects between genotypes within a population were typically small relative to the overall gains in fitness measured against the ancestor (Elena & Lenski, 1997; Maddamsetti et al., 2015). Therefore, the frequency-dependent effects could be ignored in most populations when interpreting fitness trajectories measured over long periods of evolution (Lenski et al., 2015; Wiser et al., 2013).

6.1.3. *Spontaneous mutation*

The LTEE depends on spontaneous mutations to provide the genetic variation that fuels adaptation by natural selection. Most evolutionary biologists accept this approach, but others with more applied interests have sometimes asked whether the bacteria should be mutagenized to accelerate their evolution. There might be some experiments for which increasing the mutation rate would be useful, but it is often unnecessary and can lead to complications. First, as explained above, experiments that preceded the LTEE showed that fitness gains would occur in a reasonable timeframe. Second, the speed of adaptation in large asexual populations is generally not strongly limited by the supply of new mutations, owing to clonal interference among lineages with different beneficial alleles that must compete with one another (de Visser et al., 1999; Gerrish & Lenski, 1998; Izutsu et al., 2024; Wiser et al., 2013). Third, the fitness-impairing effects of deleterious mutations will be amplified if one increases the mutation rate. Fourth, a mutagenic procedure adds complexity to the workflow and increases the potential for mistakes. Moreover, one must consider potential safety and environmental issues with some mutagens. For projects where it is deemed important to increase the mutation rate, one should consider using strains with defects in DNA repair that cause hypermutability as an alternative to mutagens.

Fifth, and perhaps most importantly when designing a future experiment, increasing the mutation rate will increase the number of "passenger" mutations (i.e., neutral and weakly deleterious hitchhikers). That, in turn, will make it more difficult to identify the beneficial "driver" mutations that are responsible for adaptation. Gene-level parallelism provides a powerful signature for identifying beneficial mutations in the LTEE, but that parallelism is obscured by widespread hitchhiking in those populations that evolved hypermutability (Tenaillon et al., 2016).

6.2. Major Findings from the LTEE

6.2.1. *Repeatability of evolution*

The LTEE has 12 replicate populations, all derived from the same ancestral strain and propagated in the same conditions. This replication allows one to examine the repeatability of evolution. Evolutionary change is driven both by the determinism of natural selection and the random forces of mutation and genetic drift; even beneficial mutations are often lost by chance while they are rare. Over the course of the LTEE, many examples have been documented of parallel (i.e., repeatable) changes across many or all of the replicate populations as well as idiosyncratic changes that are unique to a particular population (Blount et al., 2018; Lenski, 2017a). For example, parallelism was shown by the accrual of beneficial mutations in many of the same genes across most or all of the replicate populations (Tenaillon et al., 2016; Woods et al., 2006). Only rarely, however, did the exact same mutation arise in even two populations (Maddamsetti et al., 2017; Woods et al., 2006). Also, only one population has evolved the ability to consume citrate, as we will describe in more detail below. A sort of middle ground, in terms of repeatability, is the fact that half the populations evolved hypermutability, which caused them to accumulate point mutations at a much higher rate than the other populations (Sniegowski et al., 1997; Tenaillon et al., 2016; Wielgoss et al., 2013).

Despite the differences in mutation rates, most populations have evolved similar fitness levels under the conditions of the experiment, albeit with small and persistent differences in their fitness trajectories (Lenski et al., 2015). Those small differences suggest that the populations have been ascending subtly different fitness peaks or ridges in the *adaptive landscape* — in other words, finding similar but not identical adaptive solutions to the challenges imposed by the conditions of the LTEE. And while most populations have similar fitness levels in the LTEE environment where they evolved, they have diverged to a much greater extent in their fitness when faced with different resources and various stresses (Card et al., 2021; Cooper et al., 2001; Cooper & Lenski, 2000; Leiby & Marx, 2014; Travisano & Lenski, 1996). The replicate populations have also evolved similar increases in their average cell size (volume), but with considerable divergence in cell shapes (Grant et al., 2021). Finally, many populations evolved negative frequency-dependent interactions that have allowed coexistence of two or more lineages based on cross-feeding of metabolic byproducts (Elena & Lenski, 1997; Good et al., 2017; Großkopf et al., 2016; Rozen & Lenski, 2000; Turner et al., 2023). On balance, then, the replicate populations seemed to have followed similar but subtly different evolutionary paths, with one very conspicuous exception, which we discuss next.

6.2.2. *Evolution of the novel ability to consume citrate*

The most striking change that has occurred during the LTEE is that, at about 31,000 generations, a lineage in one population evolved a novel ability to consume the citrate in the DM25 medium (Blount et al., 2008, 2012). Even after 80,000 generations, none of the other 11 populations has evolved this capability. Moreover, as a species, *E. coli* is characterized by its inability to grow on citrate in the presence of oxygen (Scheutz & Strockbine, 2005). Some *E. coli* strains can consume citrate under anoxic conditions provided that another energy source, such as glucose, is also available; however, the LTEE ancestor cannot do even that. Over the years, some *E. coli* strains have been found in nature that are able to grow aerobically on citrate. These atypical strains evidently acquired that ability via horizontal transfer of plasmids from other species (Ishiguro et al., 1979). However, there are no such opportunities for acquiring plasmids in the LTEE.

The ability to consume citrate gave the bacteria a clear advantage by providing access to an abundant but unused source of carbon and energy. As a consequence, the citrate-using lineage reaches a substantially higher density than the other LTEE populations. Why then was it so difficult to evolve this new function? In brief, there are two plausible explanations (Blount et al., 2008). According to one hypothesis, the ability to use citrate required an unusual type of mutation. Consider the alternative possibility that a single point mutation would suffice to gain this new ability. Given the size of the populations, the rate of point mutations (even for those that did not evolve hypermutability), and the number of generations, all of them should have had any single mutation and evolved this advantageous ability. But perhaps a much rarer type of mutation was required — say, an inversion of some particular region of the genome involving two specific endpoints. The second hypothesis invokes *historical contingency*, such that the effect of the mutation that gave rise to the ability to consume citrate depended on one or more prior mutations. As shown by Zachary Blount, both hypotheses are correct, and together they explain the difficulty of evolving this new capability. The mutation that produced the first citrate-using cell was an unusual tandem duplication that produced a new arrangement of regulatory and structural sequences, and it allowed the efficient use of citrate only in conjunction with certain other mutations (Blount et al., 2008, 2012; Quandt et al., 2014, 2015).

6.2.3. *The dynamics of adaptive evolution*

Recall that the second set of questions addressed by the LTEE concerns the tempo and mode of evolutionary change. Is evolution always gradual, or are there periods of faster and slower evolution? And will fitness eventually reach some upper limit, or can it continue indefinitely even in a constant environment? In sexually reproducing populations that start with abundant genetic variation, phenotypic evolution tends to be gradual (Falconer & Mackay, 1996; Walsh & Lynch, 2018). In asexual populations that begin without standing variation, by contrast, phenotypic evolution requires that new mutations occur that confer beneficial effects. Moreover, it takes time for those new mutations to spread through a large population.

For example, in the LTEE, the most beneficial mutations that occur (leaving aside the case of citrate use) provide a growth-rate advantage of roughly 10% (Barrick et al., 2009; Lenski et al., 1991). It takes about 250 generations for a mutant with a 10% advantage to increase from a single individual to become the majority of a population of 500 million cells (Lenski et al., 1991). And because growth rates — and differences in growth rates — compound exponentially, for most of that time the beneficial mutant is so rare that it has no appreciable effect on a population's average fitness. But between 200 and 300 generations, the mutant goes from being a small minority to the vast majority, resulting in an almost step-like increase in average fitness (Figure 6.1). The same step-like changes also occur for phenotypes correlated with fitness, such as cell size in the case of the LTEE (Elena et al., 1996). Over time, the LTEE's step-like fitness gains have become progressively smaller, and the changes appear more gradual. Nonetheless, the ongoing fitness gains are still driven by beneficial mutants that begin as single cells and then spread exponentially. Thus, the step-like changes continue, but as the steps

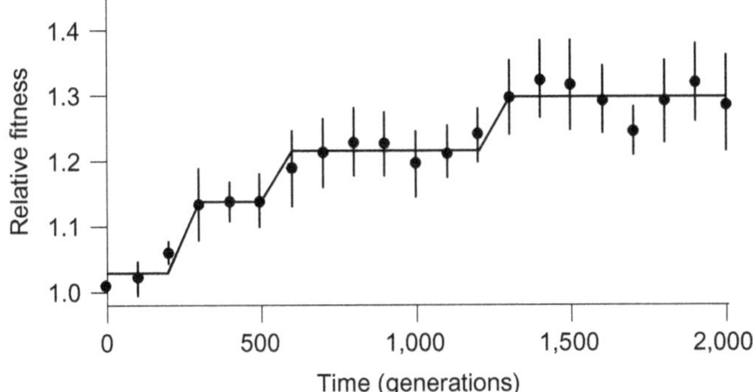

Figure 6.1: Step-like increase in fitness during the earlier generations of the LTEE. Symbols show fitness values measured at 100-generation intervals from one LTEE population. Error bars show 95% confidence limits. The solid line shows the fit of a step model to these data. Lake et al. (2025) present a software program that recapitulates and explains the step-like dynamics seen in the LTEE (adapted from Figure 1 in Lenski [2017c]).

become smaller and more spread out over time, they become difficult to distinguish from the measurement noise associated with estimating fitness, giving the appearance of gradual change (Jagdish, 2023).

The declining rate of fitness improvement over time indicates a phenomenon called *diminishing-returns epistasis* (Figure 6.2, panel a), in which the fitness gain of a particular beneficial mutation tends to be smaller when it occurs in a more-fit genetic background. Given the declining rate of improvement, it might seem reasonable to suppose that the populations will reach some fitness peak, or upper limit, after which no more gains are possible. Alternatively, perhaps there are ever smaller steps that can proceed indefinitely far into the future. To investigate that question, Wiser et al. (2013) performed competitions to measure the fitness of samples taken at many timepoints from each LTEE population relative to the common ancestor. They then compared the trajectories to two simple models, each with only two parameters: one is called a rectangular hyperbola, and the other a power law. The hyperbolic model has an asymptote (i.e., an upper bound), while the power law does not, even though the rate of fitness gain becomes slower and slower over time. Both models provide an excellent fit to the measured trajectories of relative fitness, although the power law provides a significantly better fit. More

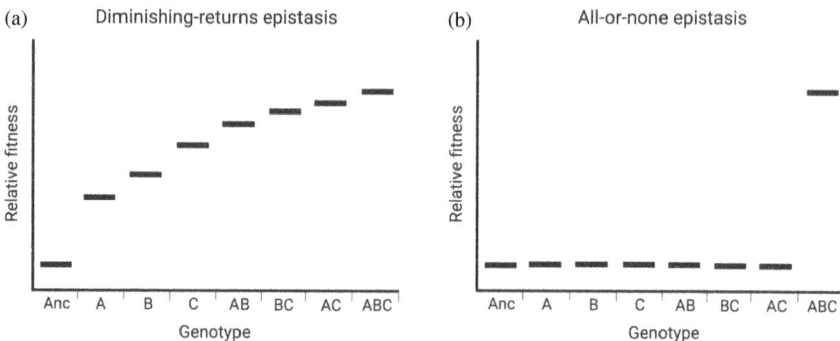

Figure 6.2:   Two of the many possible types of interactions, or epistasis, between mutations. (a) Diminishing-returns epistasis, in which the combined effect of multiple beneficial
mutations is smaller than expected from their individual effects. This type of epistasis is
common in the LTEE. (b) All-or-none epistasis, in which some beneficial effect requires the
presence of multiple mutations. This type of epistasis was observed in an experiment with
coevolving phage lambda and bacteria (adapted from Box 2 in Barrick & Lenski [2013]).

importantly, the power law accurately predicts fitness gains, whereas the
hyperbolic model fails in that regard (Figure 6.3). In fact, Wiser et al.
(2013) showed that the power law could accurately predict fitness gains
out to 50,000 generations using only the data collected from the first
5,000 generations. By contrast, the hyperbolic model consistently under-
estimated future gains. Later work extended the empirical data and model
projections to 60,000 generations, showing that fitness continued to
increase slowly, at a rate consistent with the prediction based on the
power law fit to earlier generations (Lenski et al., 2015). Both studies
also found that the LTEE populations that evolved hypermutability
achieved slightly higher fitness than those that retained the low ancestral
mutation rate (Lenski et al., 2015; Wiser et al., 2013). It is also notewor-
thy that the genetic targets of selection have changed over time in the
LTEE, owing to epistasis that alters the strength and, in some cases, the
direction of selection acting on mutations in various genes (Couce et al.,
2024; Good et al., 2017). Taken together, these results demonstrate that
adaptation to a simple and constant environment is more complex and
dynamic than often assumed.

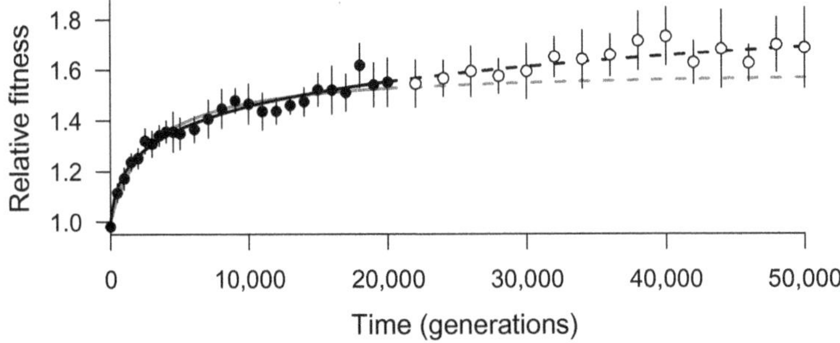

Figure 6.3: Two models predict different fitness trajectories in the LTEE. Relative fitness was estimated for most LTEE populations through 50,000 generations. The population that evolved to use citrate is not included in this analysis for timepoints after it evolved that novel ability. The symbols show the average fitness of the replicate populations; the error bars show 95% confidence intervals. The solid black and solid grey curves show the best fits of the power-law and hyperbolic models, respectively, to the data from the first 20,000 generations only. The dashed black and dashed grey curves show the predicted trajectories through 50,000 generations based on the best fits to the early data of the power-law and hyperbolic models, respectively. Both models provide good fits to the observed data, but the power-law model does a much better job at predicting the future trajectory (adapted from Figure 2B in Wiser et al. [2013]).

6.2.4. *Coupling of phenotypic and genomic evolution*

Given that new mutations are constantly occurring, one might imagine that genome evolution proceeds at a more or less constant rate, as suggested by the idea of a molecular clock. However, the LTEE paints a more complex and nuanced picture. First, it is important to realize that genomic change in the LTEE depends on a complex interplay of three evolutionary processes: mutation, selection, and genetic drift. It is sometimes wrongly presumed that random drift is important only in the absence of selection. In fact, drift occurs with or without selection. When drift is combined with selection, it leads to hitchhiking. This effect is sometimes called "genetic draft" because a hitchhiking mutation can be pulled to high frequency in the same way that one cyclist can benefit by drafting behind another. Second, the rate parameters that govern mutation and selection have changed dramatically over the course of the LTEE. As discussed in the section above, the rate of fitness improvement has declined over time.

As also mentioned already, some of the populations evolved hypermutability, whereas others retained the low ancestral mutation rate (Figure 5.2, compare top and bottom panels). The effects of both of these changes are clearly seen in the genomic data, and we will explain them in turn.

To understand the effect of the decelerating rate of adaptation on the tempo and mode of genome evolution, we first consider only the lineages that retained the low ancestral mutation rate. The rate at which beneficial mutations accumulate in the genomes declined over time, as mutations that confer smaller benefits require longer to spread to fixation. However, neutral mutations should continue to accumulate at the same expected rate: any particular neutral mutation might be promoted or impeded by a *selective sweep*, depending on whether the neutral and beneficial mutations occur in the same or different lineages, but on average the expected rate of accumulation of neutral mutations is constant. Thus, the overall rate of genomic change is the sum of a constant rate plus a rate that reflects the decelerating fitness trajectory, giving a more nearly constant slope for the genomic trajectory than for the fitness trajectory (Barrick et al., 2009; Tenaillon et al., 2016).

In addition, the ratio of nonsynonymous to synonymous mutations started out extremely high in the LTEE and then has declined over time (Tenaillon et al., 2016; see Figure 6.4). Recall that nonsynonymous changes are point mutations in protein-coding sequences that cause the substitution of one amino acid for another in the encoded protein. Synonymous changes, by contrast, do not cause any change in the protein owing to the redundancy of the genetic code. Any particular nonsynonymous mutation is beneficial, neutral, or deleterious, depending on the effect of the altered amino acid sequence, whereas the vast majority of synonymous changes are neutral because they do not affect the protein's structure. Synonymous changes are thus widely used as a benchmark for the rate of molecular evolution in the absence of selection. The proportion of nonsynonymous mutations in the LTEE was initially so high because the fastest way for a mutation to spread is by natural selection. Over time, however, the beneficial mutations tended to have smaller effects and took longer to spread, while neutral mutations continued to accumulate at the same expected rate. As a result, the ratio of nonsynonymous to synonymous mutations has steadily declined over time in the LTEE (Figure 6.4).

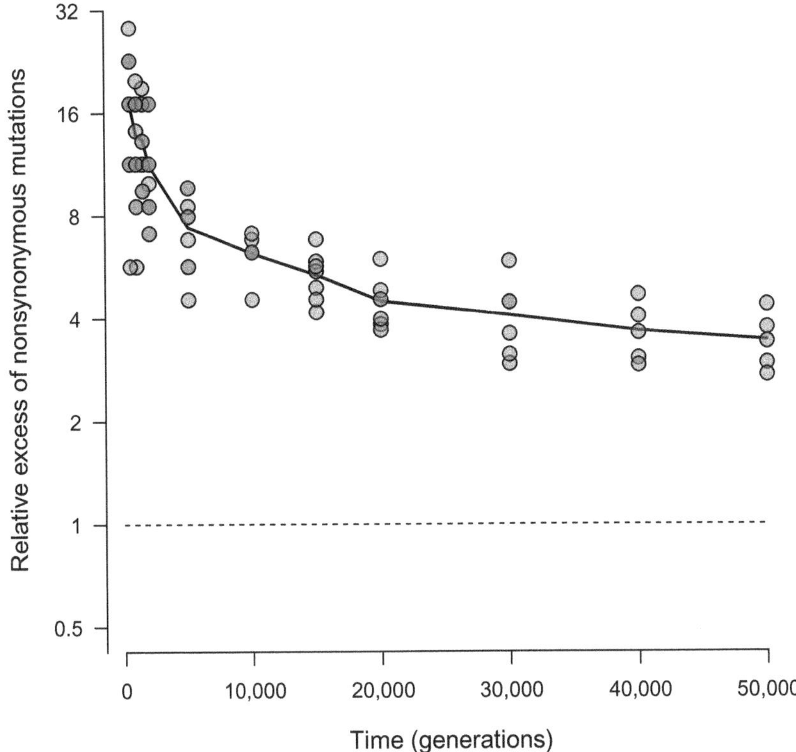

Figure 6.4: LTEE populations show an excess of nonsynonymous relative to synonymous substitutions. Symbols show values for the replicate populations obtained from whole-genome sequences, excluding later generations from those populations that had evolved hypermutability. The solid line shows the average of the replicate populations. Nonsynonymous point mutations change the amino acid sequence of an encoded protein, whereas synonymous mutations do not change the protein owing to redundancies in the genetic code. Synonymous substitutions are therefore often used as a benchmark for selectively neutral genomic evolution. An excess of nonsynonymous substitutions, as seen in the LTEE, indicates that genomic evolution is largely driven by beneficial mutations that sweep through the evolving populations (adapted from Figure 4b in Tenaillon et al. [2016]).

The overall rate of genomic change and the ratio of nonsynonymous to synonymous mutations were both strongly affected by the evolution of hypermutability in some LTEE populations. While those populations saw only a modest increase in their rate of fitness gains, they experienced a dramatic increase in the rate at which mutations accumulated (Tenaillon et al., 2016; Wielgoss et al., 2013). Moreover, the specific types of mutations they accumulated depended on the mutations that caused the hypermutability. For example, some hypermutable strains had defects in the proofreading that occurs during DNA replication, and these strains are biased toward *transition* mutations. Transition mutations occur when one purine (adenine or guanine) replaces the alternative purine, or when one pyrimidine (cytosine or thymine) replaces its counterpart. Other hypermutable strains fail to remove certain oxidized nucleotides that can be mis-incorporated into DNA while it is being replicated, and these defects cause particular *transversion* mutations. Regardless of the specific type of mutator strain, the increase in the mutation rate led to only a small increase in the number of beneficial driver mutations but a much larger increase in the number of neutral or nearly neutral hitchhiking passenger mutations. Consequently, the ratio of nonsynonymous to synonymous substitutions is lower in the hypermutable lineages than in those with the low ancestral mutation rate. The increased number of passenger mutations also makes it much harder to use parallelism to identify the beneficial mutations that produced the fitness gains. Over time, most lineages that became hypermutable either reverted to the low ancestral mutation rate or evolved compensatory changes that led to an intermediate mutation rate (Tenaillon et al., 2016; Wielgoss et al., 2013).

Tenaillon et al. (2016) also compared patterns of genome evolution in the LTEE populations to *mutation-accumulation lines* that were repeatedly propagated as single colonies, with each colony derived from a single cell (recall Figure 3.1, panel c). These severe bottlenecks eliminate genetic variation and thus minimize the efficacy of natural selection. Compared to these mutation-accumulation lines, the LTEE populations that retained the low ancestral mutation rate had excesses of nonsynonymous mutations, intergenic mutations, insertions, and deletions. This comparison further supported the conclusion that many or most of the mutations that reached high frequency in the non-hypermutable LTEE populations were beneficial drivers.

Sequencing also revealed that genetically divergent lineages have coexisted for thousands, and even tens of thousands, of generations in some of the LTEE populations (Good et al., 2017). Such sustained coexistence implies the action of negative frequency-dependent selection, in which each of two distinct ecotypes has a fitness advantage when rare. Coexisting lineages had been observed previously in the LTEE (Le Gac et al., 2012; Rozen & Lenski, 2000; Rozen et al., 2005), but intensive sequencing showed that such coexistence was more common than expected given the simplicity of the LTEE environment. However, the coexistence can be disrupted if a beneficial mutation occurs in one ecotypic lineage that is so advantageous that it not only dominates its own lineage but also overcomes the advantage when rare of the alternative ecotype (Maddamsetti et al., 2015). When that happens, many linked mutations can sweep to fixation simultaneously (Good et al., 2017; Maddamsetti et al., 2015).

By 50,000 generations, the average genome length in the LTEE had declined by ~1.4% relative to the ancestor, as deletions tended to be larger than insertions and more frequent than duplications (Tenaillon et al., 2016). The reduction in genome length occurred even though individual cells had become larger (Grant et al., 2021). By contrast, genome length and cell size tend to be positively correlated across prokaryotic species (DeLong et al., 2010; Marshall et al., 2022). The LTEE populations also evolved increased levels of mRNA per cell, which scaled with the increase in cell size (Favate et al., 2022).

6.3. Beyond the LTEE

6.3.1. *Experimental evolution in asexual yeast*

Michael Desai and his team have run the largest and longest evolution experiment with asexual yeast to date (Johnson et al., 2021). Their experiment encompassed 205 populations of the yeast, *Saccharomyces cerevisiae*, that evolved for ~10,000 generations in three environments with different culture media and temperatures. In each environment, some of the founding strains were haploid (with two different mating types), while others were diploid. The size of this study was made possible by new technology that did not exist at the start of the LTEE — namely, they used a robotic system to perform the daily serial transfers. The study began with 270 populations, but 65 were lost to contamination and other errors. Johnson et al. (2021) ran competition assays at multiple generations to measure changes in relative fitness. They also performed whole-population, whole-genome sequencing at six timepoints for 90 of the populations.

Similar to the LTEE, Johnson et al. (2021) observed declining rates of fitness improvement over the course of their experiment as well as the steady accumulation of mutations, widespread genetic parallelism, and some historical contingency. These patterns were seen in both the haploid and diploid yeast. They also saw evidence of clonal interference, similar to that observed in the *E. coli* LTEE. Unlike the LTEE, however, Johnson et al. (2021) did not find evidence of stably coexisting lineages in the yeast populations. They suggested that coexisting lineages might have arisen but gone extinct between the timepoints when they sequenced samples. Even so, their results ruled out the long-term coexistence of lineages as occurred in the LTEE (Good et al., 2017; Rozen & Lenski, 2000) and some other experiments (Behringer et al., 2018).

Johnson et al. (2021) did not observe the evolution of hypermutable lineages, again unlike the LTEE (Sniegowski et al., 1997; Tenaillon et al., 2016). They suggested that this difference might reflect the initial fitness of the ancestral strains and the resulting balance between beneficial and deleterious mutations in the experimental environments. Alternatively, they suggested hypermutator strains may not have arisen in their study because the ancestral strains had higher mutation rates than the LTEE ancestor. However, Johnson et al. (2021) did observe some differences in the rate of mutation accumulation depending on the ancestral mating type and ploidy level. They also observed that the ratio of nonsynonymous to synonymous mutations was close to one, suggesting that selection for beneficial nonsynonymous mutations was roughly offset by hitchhiking of neutral synonymous mutations and selection against deleterious mutations.

In their yeast experiment, Johnson et al. (2021) observed extensive genetic parallelism, similar to the LTEE. Parallel changes occurred most often in certain biosynthetic and signaling pathways. A particularly striking case was the evolution of adaptations to compensate for a deleterious mutation that was present in the ancestral strain, which had a premature stop codon in the ADE2 gene that disrupts the adenine biosynthesis pathway. Adenine depletion can limit growth even in rich media, and the mutation in ADE2 also caused a toxic intermediate to accumulate. As a consequence, loss-of-function mutations upstream of this pathway, which are deleterious when ADE2 is functional, became beneficial because they prevented the buildup of the toxic compound. Indeed, Johnson et al. (2021) found loss-of-function mutations in the ADE pathway, usually upstream of ADE2, in almost every sequenced population. They also noted that these mutations provide an example of historical contingency, whereby future evolution depends on the specific genetic background, similar to that observed in LTEE.

Another interesting feature of this yeast experiment was comparing evolution in asexual haploids and diploids. During their experiment, one haploid population underwent a whole-genome duplication, thus becoming diploid. Although it occurred in only one population in this study, other experiments with haploid yeast have documented many such events (Harari et al., 2018). More generally, an important difference between haploids and diploids involves the *dominance* effects of mutations (Marek & Korona, 2016). Some mutations that are beneficial in haploids will be recessive or only partially dominant in diploids. If fully recessive, they will be invisible to selection when they first arise in a diploid cell; even those that are partially dominant will often be less beneficial in a diploid background. Consistent with past experiments, Johnson et al. (2021) found that diploid populations had slower rates of fitness gain than haploids. However, the diploid populations did not accumulate mutations at a lower rate than the haploids. Another difference between the haploid and diploid populations was that some recessive deleterious mutations accumulated in the diploids, presumably by hitchhiking. In fact, many of the deleterious mutations were in essential genes.

6.3.2. *Modes of reproduction and genetic diversity*

Evolution experiments with yeast can also be run to compare the effects of sexual and asexual reproduction. One can change the mode of reproduction in yeast either genetically or by manipulating the culture environment. By evolving yeast with contrasting modes of reproduction, some studies have tested the effect of sex on the rate of adaptation (Goddard et al., 2005; Zeyl & Bell, 1997). However, no clear consensus has been reached to date, and more studies would be worthwhile to probe this big question.

Another set of questions concerns the dynamics of genome evolution in experimental populations of asexual microbes and how these dynamics compare with sexually reproducing multicellular animals. In particular, there are several distinguishing features of these experimental systems that are hard to disentangle. Most evolution studies with asexual microbes begin without any standing genetic variation, population sizes are large (typically millions of individuals in each replicate), and experiments run for many generations (usually hundreds and often thousands). By contrast, most evolution studies with sexually reproducing animals start with abundant genetic variation, population sizes are relatively small, and the duration in generations is comparatively short.

Only a few experiments to date have sought to disentangle some of these confounding factors. As mentioned in Chapter 5, Souza et al. (1997) compared the rates of adaptive evolution in strictly asexual bacterial populations with other populations that had opportunities for plasmid-mediated acquisition of genetic variants from a different strain of *E. coli*. However, they saw no significant difference between the two treatments. Later genomic analysis indicated that the benefit of recombination, if any, was offset by the genetic load caused by replacing previously evolved beneficial mutations with maladapted alleles (Maddamsetti & Lenski, 2018). Burke et al. (2014) ran an experiment with populations of yeast obtained by recombination between several strains in order to produce a high initial level of genetic variation. The populations were propagated in a manner that also forced periodic mating and recombination. Even with large populations and after hundreds of generations, new mutations had contributed little to adaptation. Instead, they found that the initial variation had fueled the observed evolution. It would be interesting if, in the future, a similar experiment with yeast also included asexual populations derived from the same ancestor and evolving in parallel. In an experiment with strictly asexual bacteria, Izutsu and Lenski (2022) compared the rates of adaptation to a new medium in populations propagated from single cells and others that began by mixing several clones or even entire populations from the LTEE. The initial diversity influenced the dynamics early in the experiment. By 500 generations, however, the populations that relied entirely on new mutations had reached the same fitness level as those that began with substantial genetic variation.

6.3.3. *Asexual versus sexual evolution in viruses*

Paul Turner and Lin Chao (1998, 1999) performed experiments with phage φ6 and its bacterial host *Pseudomonas phaseolicola* in order to compare the rate of adaptation between asexual and sexual populations of that virus. Viral genomes can undergo genetic recombination only when two or more viruses coinfect the same cell. The φ6 genome has three physically disjoint segments, and recombination occurs when different segments derived from coinfecting viruses are packaged in the same viral capsule during infection. By manipulating the ratio of viruses to host cells (called the multiplicity of infection, or MOI), Turner and Chao were able to vary the potential for recombination. They predicted that a high MOI would increase adaptation by allowing independently arising beneficial mutations to recombine and spread simultaneously, rather than sequentially, through the virus populations (Maynard Smith, 1978; Muller, 1932). To test that prediction, they performed paired-growth assays to measure the relative fitness of the evolved and ancestral viruses. These assays were performed in such a way that the evolved and ancestral viruses could not coinfect the same individual cells.

To their surprise, however, Turner and Chao (1998, 1999) discovered that the viruses that evolved in the high MOI treatment, and which thus had the opportunity for recombination, were significantly less fit than those evolved in the low MOI treatment, where all infections resulted from single viruses. Why this unexpected result? Turner and Chao reasoned that the infections that resulted from multiple viruses had another effect, besides allowing recombination. In particular, if two or more viruses infected the same host cell, then they would compete for intracellular resources. For example, if the ancestral virus split its resources evenly between replicating its genome and making the proteins used to build the capsid (required for virus transmission), while a mutant virus devoted all of its resources to copying its genome, then the mutant would have a within-host competitive advantage. Turner and Chao dubbed the two viral strategies "cooperators" and "defectors," respectively.

To test this hypothesis, Turner and Chao (1998, 1999) ran additional fitness assays using a different design, one that better matched the difference between the high and low MOI treatments in the evolution experiment itself. To that end, they mixed the evolved and ancestral viruses before exposing them to host cells. With this assay design, coinfections of the evolved and ancestral viruses would be common at high MOI, but rare at low MOI. When these assays were performed at low MOI, the viruses that evolved in the low MOI treatment were still more fit than those that evolved at high MOI. However, when the assays were run at high MOI, the viruses that evolved at high MOI were more fit than had been measured in the paired-growth assays or when measured at low MOI. Thus, the paired-growth assays did not account for the within-host competition, which was an important factor driving the evolution of the viruses in the high MOI treatment.

The work by Turner and Chao illustrates a very important aspect of experimental design — namely, a manipulation intended to affect one process (in this case, genetic recombination) may inadvertently also influence another process (here, within-host competition). One should be mindful of this possibility when designing evolution experiments and, like these authors, carefully consider and rigorously test alternative hypotheses.

6.3.4. *Evolution of a new viral function*

In evolution experiments, just as in nature, adaptation typically occurs by improving existing functions — for example, faster growth or more efficient resource use. Sometimes, however, new capabilities emerge by modifying and co-opting existing structures and functions and using them in novel ways (Darwin, 1859; Jacob, 1977). Earlier in this chapter, we discussed an example from the LTEE, namely the ability to use citrate that arose in one population (Blount et al., 2008, 2012). Another example was discovered in evolution experiments with phage λ, which gained the ability to infect *E. coli* through a receptor on the cell surface that could not be used by the ancestral virus (Meyer et al., 2012). The authors sought to identify the specific mutations that conferred the new function and to understand the evolutionary forces that promoted its emergence.

Phages initiate infections by binding to specific receptors on the surface of their bacterial hosts and then inject their genomes and thereby commandeer the infected cells to make more viruses. Bacteria may evolve resistance to phage infection in various ways, such as by reducing the production of the targeted receptor or altering its structure. Phage λ has a specialized ligand, called the J protein, at the end of its "tail" that targets a protein, called LamB, on the surface of the bacteria. LamB is used by *E. coli* as a porin that allows maltose and other glucose polymers to diffuse across the outer membrane. For phage λ, LamB is its only mode of entry into the cell.

Meyer et al. (2012) cultured six replicate communities containing phage λ and *E. coli* for 28 days with daily dilutions, and samples were collected weekly. Glucose provided carbon and energy for the bacteria. Previous work had shown that these conditions favored mutations in the bacteria's *malT* gene, which encodes a transcription factor that promotes expression of the LamB protein targeted by λ. By reducing the expression of LamB, these mutations provide the bacteria with partial protection against infection.

Meyer et al. (2012) tested the capacity of the evolved phage to infect cells via a new receptor by plating them on lawns of an *E. coli* mutant that is missing the *lamB* gene and cannot produce the LamB receptor. One of the evolved λ, which they called EvoC, was able to infect that mutant. That meant that EvoC had gained the ability to use a receptor that its ancestor could not use. To identify the novel receptor, they tested the ability of EvoC to infect a set of *E. coli* mutants that were missing various outer membrane proteins. The only host strain resistant to EvoC was a double mutant that lacked expression of both LamB and a protein called OmpF. This result showed that OmpF was the new receptor and, moreover, that EvoC could also still use the LamB receptor. OmpF and LamB are both porins with trimeric structures, although their amino acid sequences are highly diverged. The J protein at the end of λ's tail is also a trimer, which suggests that a trimeric structure is an important feature for its recognizing and binding to host receptors.

Meyer et al. (2012) then examined the mutations that conferred the novel ability to use the OmpF receptor. They sequenced the EvoC genome and found five mutations, all in the gene that encodes the J protein. Another phage isolated from the same population that could not infect via OmpF differed from EvoC by just one of those five mutations, indicating that it played a critical role in gaining the new function. But was that single mutation sufficient to produce the new ability?

To answer that question and better understand how λ evolved to use this new receptor, Meyer et al. (2012) ran an additional 96 replicates of their experiment with coevolving phage and bacteria. In 24 cases, λ evolved the ability to use OmpF. They then sequenced the gene encoding the J protein from these 24 replicates along with 24 replicates that did not evolve that ability. They found a total of 241 mutations across the 48 J alleles, all of them nonsynonymous point mutations. They also observed striking parallelism, as there were only 40 unique mutations. Two mutations were identical in all 24 phage isolates that evolved the new ability to use OmpF, and two other sets of almost identical mutations (i.e., affecting the same or nearby codons) were also found in those 24 isolates (Figure 6.5). None of the 24 phage isolates that could only use the original LamB receptor had all four of the mutations that were evidently required for λ to target OmpF.

Thus, an "all or none" interaction among these mutations appeared to be responsible for the evolved ability to use the OmpF receptor (Figure 6.2, panel b). Further work confirmed that the phage could not use that receptor if any of the four mutations were absent (Meyer et al., 2012). That raised another question: How did the four mutations accumulate sequentially if all of them were necessary for the new functionality? The authors argued that each of the mutations was beneficial, even before they collectively enabled the virus to use the OmpF receptor, based on three lines of evidence. First, all of the observed mutations in the J alleles were nonsynonymous, which could not be explained by chance alone and implies selection favored those mutations. Second, the mutations were highly concentrated in the region of the J protein that interacts directly with the cell's LamB protein. Third, such parallelism is a hallmark of adaptive evolution. The authors thus concluded that the individual mutations likely improved the interaction of λ's tail with the original LamB receptor, which may have been especially important after the bacteria evolved reduced expression of LamB. Indeed, a further experiment showed that λ's ability to use OmpF did not evolve when the bacteria had mutations that blocked a different step in the infection process (Meyer et al., 2012).

In summary, this study illustrates some of the challenges of using experimental evolution to study the origin of new functions: the need for multiple mutations; complex interactions among mutations, in this case

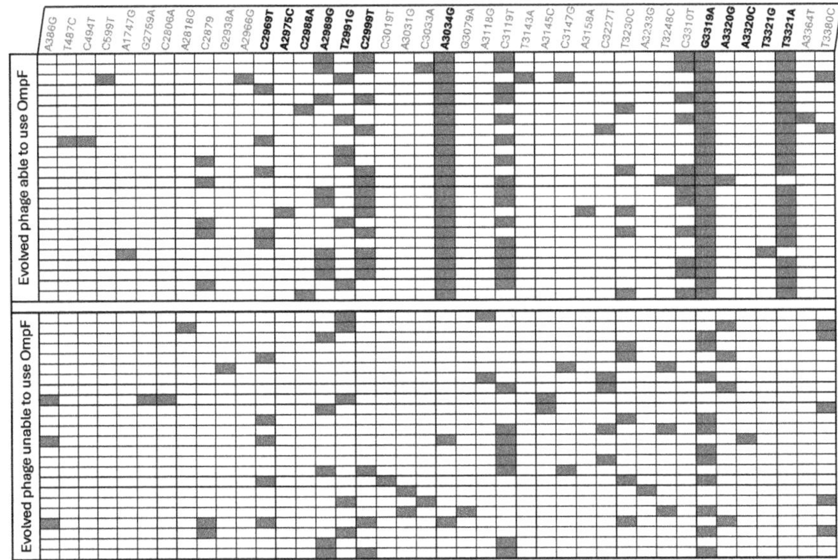

Figure 6.5: Lambda phage that evolved the new ability to infect host cells via the OmpF receptor possess a set of four mutations that are not present in those that remain dependent on the ancestral LamB receptor. The gene encoding the J protein that interacts with receptors on the bacterial cell surface was sequenced in 48 phage that had coevolved with *E. coli*. That gene is over 3,000 bp in length; only those positions that changed in at least one evolved phage are shown along the top of this figure. Gray cells indicate mutations in an evolved phage. The 24 phage that evolved the ability to use the OmpF receptor are shown first, followed by 24 evolved phage that could still use only the ancestral LamB receptor. All 24 evolved phage that can use the OmpF receptor have similar sets of four mutations, including two specific changes present in all of them as well as two other changes that vary slightly but involve nearby positions. None of the 24 evolved phage that cannot use OmpF have all four of these mutations. The essential mutations are shown in black along the top, while all other mutations are shown in a grey font (adapted from Figure 3 in Meyer et al. [2012]).

both within and between species; and historical contingency, such that phenotypic changes depend on prior mutations. Justin Meyer and colleagues later studied the role of host evolution in λ's ability to use the new receptor, the biophysical changes in the J protein during the transition, and the impact of this key innovation on viral recombination and speciation (Burmeister et al., 2016; Gupta et al., 2022; Meyer et al., 2016; Petrie et al., 2018; Strobel et al., 2024).

6.3.5. *Other exemplary evolution experiments using asexual microbes*

It is not feasible to discuss all of the fascinating evolution experiments that have been performed using asexual microorganisms in a primer. However, we will now briefly call attention to three other beautiful experiments that interested readers can explore further on their own.

Paul Rainey and Mike Travisano (1998) compared the evolution of a bacterium, *Pseudomonas fluorescens*, in two environments. In one environment, the cultures were constantly shaken, while in the other they were static; the two environments were otherwise identical. The authors found that the bacteria quickly diversified into three distinct ecotypes in the static cultures but not in the cultures that were shaken, as a consequence of the physical structure and chemical gradients that emerged in the static environment. Rainey and colleagues further investigated the ecological factors as well as the genetic and phenotypic changes responsible for this diversification (Fukami et al., 2007; Gallie et al., 2019; Kassen et al., 2000; Lind et al., 2019; McDonald et al., 2009; Spiers et al., 2002).

Greg Velicer et al. (2000) studied *Myxococcus xanthus*, a bacterium that exhibits complex social behaviors, including the formation of multicellular fruiting bodies under stressful conditions. Most cells that engage in this cooperative process die, but some form spores that can survive adverse conditions. The authors reasoned that this system might be vulnerable to cheaters, and they demonstrated that some lineages that had evolved for many generations without selection for this cooperative behavior indeed had an advantage when mixed with their ancestors. Velicer and colleagues have continued to study the ecological and genetic factors that shape the tension between cooperation and cheating in these bacteria (Fiegna & Velicer, 2003; Fiegna et al., 2006; Manhes & Velicer, 2011; Rendueles et al., 2015; Velicer et al., 2006; Velicer & Yu, 2003; Yu et al., 2016).

As noted in Chapter 4, Will Ratcliff et al. (2012) evolved multicellular yeast from a unicellular ancestor by propagating those organisms that settled faster during centrifugation. Unlike the fruiting bodies of *M. xanthus* that form by cellular aggregation, the resulting "snowflake" yeast are produced when generations of mother and daughter cells remain physically attached to one another. Ratcliff and colleagues have continued to analyze the origin, genetics, and morphological properties of the evolved multicellular yeast (Bozdag et al., 2023; Jacobeen et al., 2018; Libby et al., 2014; Ratcliff et al., 2015; Tong et al., 2025).

6.4. Chapter Summary

The field of microbial experimental evolution has dramatically expanded as bacteria, yeast, and viruses have gone from model systems in molecular biology to favored organisms for understanding the evolutionary process. As a result, much has been learned about the repeatability of evolution, its dynamics, and how phenotypic and genomic evolution are coupled in asexual systems. The studies we have discussed also illustrate how it is important to think carefully about the organisms and environments used to address particular questions. In particular, one must consider whether results can be generalized or, alternatively, might be idiosyncratic with respect to the organism and experimental design.

Thus, one should ask a number of questions before starting a microbial evolution experiment. What are the characteristics of the ancestors? Should one begin with a single uniform strain or a diverse population? Will the populations be propagated by serial transfer, or is chemostat culture more appropriate? Is the organism amenable to freezing and revival? If so, will the frozen ancestors serve as the only controls, or will control lines be propagated in conditions that differ from the treatment populations? How frequently will samples be collected, and for how long should the experiment run? How do the culture medium and other factors relate to the research questions? Will frequency-dependent selection enhance or complicate the study? What is known about mutation rates in the study system? Can one identify hypermutators if they evolve? How will one distinguish genetic adaptation and phenotypic acclimation? Will it be possible to detect if the experiment gets contaminated with other microbes or if cross-contamination occurs between populations? Last but not least, how many replicate populations are likely needed to rigorously test one's hypotheses, and is that number practical? The answers to some of these questions will likely be tentative, and therefore you may want to perform preliminary experiments to inform your choices.

Chapter 7

Experimental Evolution in Sexual Species

Evolution experiments using sexually reproducing species have had a variety of aims, including adaptation of formerly wild organisms to lab conditions, the evolution of life-history and physiological traits, and the effects of hybridization between lineages.

7.1. Experimental Evolutionary Domestication

Studies of wild-caught organisms evolving in the laboratory have allowed biologists to observe and analyze the dynamics of evolutionary domestication (Simões et al., 2007), a topic of both theoretical interest in evolutionary biology and practical importance for conservation efforts. The adaptation of *Drosophila* populations collected from nature and then maintained in the lab has been studied extensively since the 1990s (Simões et al., 2009). Such research has illuminated basic principles of adaptation to novel environments, given that a lab is just another environment in which populations from the wild can evolve and adapt (Matos et al., 2000a; Matos et al., 2000b; Orozco-terWengel et al., 2012; Simões et al., 2019, 2017, 2009).

Some authors have argued that the use of long-established laboratory populations limits inferences about evolutionary processes (Harshman & Hoffmann, 2000; Linnen et al., 2001). This view assumes that experimental evolution studies aim to extrapolate their findings to populations in the wild. However, experimental evolution focuses primarily on testing general predictions of evolutionary theory, for which purpose any particular lab environment can serve as a novel environment to which wild-derived populations may adapt (Matos, Rego et al., 2000a). Moreover, such populations typically have abundant genetic variation, provided many founding females are included.

These studies also allow one to compare evolutionary changes in newly established lab populations with long-established lab populations to ask, for example, whether they are converging evolutionarily through time. Most evolution experiments focus on divergent evolution, analyzing the responses of long-established lab populations to novel selective regimes by comparing their state through time with that of the populations from which they derive (Garland & Rose, 2009; Kawecki et al., 2012). By contrast, in studies of *de novo* adaptation to the laboratory, the expectation is that more recently introduced populations will converge to the state of populations that have been evolving in the lab for many generations.

7.2. Case Study of Domestication in *Drosophila subobscura*

With these issues in mind, we turn to the dynamics of adaptation in populations of *D. subobscura* studied for more than 20 years in the lab of Margarida Matos. This research program has involved many experimental populations founded from natural populations that differ in their source locations and the years they were founded, providing a useful case study of evolutionary domestication.

7.2.1. *Initial studies*

This work started in the early 1990s, when it became apparent that an interesting line of research was missing in the literature: namely, how organisms sampled from wild populations evolve generation by generation after they are introduced to a laboratory environment. In particular, Matos wanted to test two theoretical expectations about adaptation to a new environment. First, the heritable variance for fitness-related traits should be depleted over time, and second, the genetic correlations between such traits would trend from positive to negative values. One inspiration came from Service and Rose (1985), who tested the second hypothesis by estimating genetic correlations in a long-established lab population of *D. melanogaster* under both its usual lab conditions and when subjected to a novel environment. In this one-generation assay, they found, as expected, a shift of genetic correlations toward positive (less negative) values under the novel conditions. But the question remained: how would such genetic correlation parameters change during prolonged adaptation to the lab?

With this question in mind, Matos et al. (2000b) followed the temporal changes in a population of *D. subobscura* recently founded by flies from natural locations near Lisbon, Portugal (Adraga). Using a half-sib mating design, the genetic structure of fecundity and longevity was analyzed every 4–6 generations for 29 generations. The study pointed to changes in the expected direction, but its low statistical power and lack of controls precluded robust conclusions. As found in other studies that combined quantitative genetics and experimental evolution (e.g., Rauser et al., 2009), the estimation of genetic variances and covariances proved to be statistically challenging.

That quantitative genetic study was complemented by characterizing evolutionary trends at the phenotypic level, after this first domesticated population had reached 24 generations. For this purpose, a new population was founded from a collection in the same location (called "W"), and this population was compared with the established population, which served as a kind of control (called "B"). Periodic assays were performed during the first 14 generations of the evolutionary domestication of the W population. That comparison lasted a bit more than one year, since the generation time in these *D. subobscura* populations was four weeks at the lab temperature of 18°C. The following traits were compared: age of first reproduction (number of days until egg laying); early fecundity (eggs laid during the first week of adult life); fecundity from days 8 to 12; and starvation resistance, which served as a surrogate for adult survival under stress generally. This study found a predictable temporal trend of improvement in fecundity traits toward the values of the established B population. But starvation resistance followed a more complex trajectory, with an initial phase of improvement to resistance greater than the B control followed by later convergence to the B level. These were interesting results, but replication and longer-established controls were still needed.

7.2.2. *Adding replication and better controls*

After the B population reached 90 generations in 1998, another new lab population was founded (called "NW") from field collections made in the same natural location. Both the B and NW populations were then split into five replicate populations after the NW population had been in the lab for two generations (Matos et al., 2002). Importantly, the founding population was maintained as an outbred population from the second generation on, with the founder females isolated in groups of around five individuals in separate vials to ensure their progeny were represented in later generations. From that point on, the five B and five NW populations were maintained as separate outbred populations, ensuring their evolutionary independence. Otherwise, culture conditions followed the same protocols, including the medium, temperature, moderate densities, discrete 4-week generations, and census population sizes of 600–1,000 individuals (Matos et al., 2002). The trajectories of adaptation of the NW populations were analyzed over the next several years (47 generations).

The big question was whether the two studies would show similar evolutionary patterns. The W and NW populations were started 6 years apart, and much might have happened in nature during that time, in addition to the effects of sampling variation. Also, the new study had replication, and the control populations were now long-established in the lab.

Three general conclusions were drawn. First, in both studies there was a clear pattern of adaptation to the laboratory environment during the early generations. Moreover, even starvation resistance showed an increase, contrary to what would be expected if populations lose functions that are important in nature due to relaxed selection in the lab (Hoffmann et al., 2001; Sgrò & Partridge, 2000).

Second, both studies showed phenotypic convergence toward the longer-established populations for traits closely related to fitness, such as early fecundity. Although convergence is commonly expected for adaptive evolution, this expectation had not yet been properly tested in the context of evolutionary domestication. Interestingly, the second study (like the first) showed a more complex trend for female starvation resistance, as the NW populations reached higher levels than the B controls in later generations. Whether convergence would eventually occur for this trait remained an open question.

Third, despite the convergence in fecundity to the controls, there were differences between the two studies, both in the level of initial differentiation and in the subsequent rates of evolutionary adaptation.

These studies suggested that, even in the same lab, evolutionary processes are sensitive to subtle differences in the founders, which may lead to important differences. They also illustrated the difficulties of real-time evolution studies in which even the controls are dynamic populations, since in *Drosophila* the ancestral state cannot be preserved by freezing to assay in synchrony with the evolving populations (in contrast with bacteria, as discussed in other chapters). Such concerns therefore raised the issue of repeatability in studies of evolutionary domestication.

7.2.3. *Examining repeatability and founder effects more broadly*

To test for repeatability and the possible role of differences in founding populations, two new sets of domestication populations were created three years later (in 2001). One set came from a collection made in Adraga (called "TW"), while the other set came from a second location (Arrábida, 50 km away, across the Tagus river) called "AR" (Simões et al., 2007). As in the previous studies, populations were replicated at the second generation in the lab. Owing to space constraints, only threefold replication was used in this study (and the one described in the next section). Analysis of these populations during their first 40 generations showed again very consistent improvement of fecundity traits in all of them, though at a slower rate for the AR populations. Also, while starvation resistance improved in the TW populations, it did not in the AR populations. Further assays of the NW populations, which had now been in the lab for around 90 generations, showed a slowing rate of improvement for life-history traits, as expected for evolutionary convergence to longer-established populations.

The different rates of adaptation for the TW and AR populations (founded by flies from Adraga and Arrábida, respectively) led to an important question: Did the difference between the two sets of lab-evolving populations reflect a difference in the wild populations from which they derived, or did it result from the random effects of sampling relatively few founders? To test these two possibilities, a new sampling design was applied in establishing new sets of evolutionary domestication experiments.

7.2.4. *Disentangling sampling effects and differences between wild populations*

In 2005, Matos and collaborators went back to Adraga and Arrábida, but this time they sampled each location twice, obtaining two independent samples from each location ("FWA" and "FWB" from Adraga, "NARA" and "NARB" from Arrábida). Then, as usual, after two generations in the lab to increase population sizes, each population was replicated threefold, creating a total of 12 new populations. Once again, the same protocols were used to generate evolutionary trajectories covering about 20 generations of laboratory domestication.

Given that the Matos lab already had data from the earlier studies for the same traits and similar generations, they addressed the question of the repeatability of phenotypic evolution and the impact of contingent factors, including differences in the initial genetic background due to geographical locations, temporal shifts, and random sampling effects.

To that end, and with the help of Larry Mueller, a broad analysis was performed of the early generations of laboratory evolution of populations derived from founders sampled in different years (1998, 2001, and 2005) and at different geographical locations (Adraga and Arrábida). With the 2005 design and data, the Matos lab could also distinguish the impact of sampling effects and geographical location (Simões et al., 2008). The analysis focused on two parameters estimated from linear trajectories over the first 15–20 generations of adaptation to the lab: the intercepts serve as surrogates for the initial differentiation between founders, and the slopes indicate the evolutionary rate. They then tested for effects across the levels of the sampling hierarchy.

The results supported the importance of contingent evolution during domestication, especially in the initial performance of populations and to a lesser extent in their subsequent evolution. Female starvation resistance showed the strongest effect of such contingency, both in initial state and adaptive rate, with founder effects across sampling years and even between the two samples from Arrábida in the same year. Fecundity traits, by contrast, showed more consistent responses, as expected given their importance for fitness.

A molecular genetic analysis of microsatellites in these populations indicated an important effect of random drift in the differences between founders (Santos et al., 2012). Although estimated population sizes in the laboratory were relatively high, the molecular markers indicated strong stochastic effects in the first two generations of domestication, leading to differentiation between founders regardless of the wild ancestral population (Santos et al., 2013). These sampling effects during the early generations of culture in the novel laboratory environment imply a substantial drop in the effective population size (N_e). Significant evolutionary changes can thus occur quickly after a founding event under domestication, with important implications for both scientific and conservation purposes.

7.2.5. *Effects of history and selection during laboratory adaptation*

The analysis of microsatellites showed that substantial differences in evolutionary dynamics reflected chance events that occurred early in lab culture and not from differences among the source populations, at least when the founders all came from nearby locations in Portugal. This finding led to a new long-term study on the impact of prior history and selection during evolution in a common laboratory environment.

D. subobscura populations exhibit substantial differentiation along their latitudinal European distribution, notably in inversion polymorphisms (Rezende et al., 2010). Taking advantage of this fact, in 2010 the Matos lab sampled populations from three widely separated European localities: Portugal (again Adraga, called "Ad"), France (Montpellier, called "Mo"), and the Netherlands (Groningen, called "Gro"). They also adjusted the initial protocol by isolating the founding females in separate vials and maintaining the first generations in the lab as families in an effort to minimize the loss of initial genetic variation and thereby reduce genetic differences between the replicate founding populations from the same location (Fragata, Simões et al., 2014). They also treated all populations with antibiotics for two generations, having detected some pathogens affecting the flies. The threefold replication of the populations was thus delayed to the fourth generation after initial sampling, but other protocols were the same as in the previous studies.

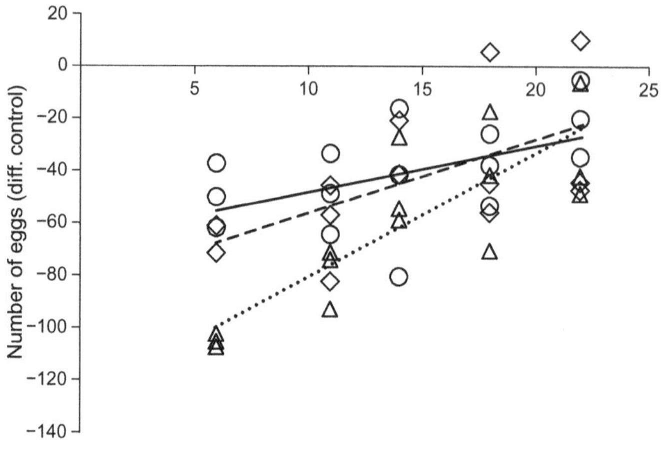

Generations

Figure 7.1:   Rapid convergent evolution of early fecundity across historically differentiated populations. The *y*-axis shows the difference in the number of eggs relative to the
control, while the *x*-axis shows the number of generations of experimental evolution.
Circles, solid line: Groningen founders; diamonds, dashed line: Adraga founders; trian-
gles, dotted line: Montpellier founders (adapted from Figure 1B in Fragata, Simões et al.
[2014]).

For life-history traits, they found that the initial differentiation
between populations was erased quickly (Figure 7.1), with populations
converging to values similar to those of the long-established TW populations in less than 15 generations (Fragata, Simões et al., 2014). However,
the frequencies of the chromosomal inversions in these populations did
not converge (Fragata, Lopes-Cunha et al., 2014).

7.2.6. *Genome-wide analysis*

Research on the genetics of the populations in the Matos lab was greatly
enhanced with the use of genome-wide sequencing (Schlötterer et al.,
2015; Turner et al., 2011), which was applied to the populations founded
in 2010 from Portugal and the Netherlands. Pool-seq was performed on
samples of 50 females per population at generations 1, 6, 25, and 50 after
lab founding, with a pooling of the TW populations used as controls in
each generation. The frequency changes of more than 3 million SNPs

were analyzed across generations, and candidate SNPs were detected by applying Cochran–Mantel tests and performing simulations (Seabra et al., 2018).

The main finding was unexpected—the phenotypic convergence between populations reported by Fragata, Simões et al. (2014) was not accompanied by any discernible loss of genome-wide differentiation (Seabra et al., 2018; see Figure 7.2). On the contrary, although Seabra et al. (2018) detected many candidate genes that appeared to have responded to selection in each population, not a single candidate SNP was in common between the populations founded from the different European locations. That finding implied that the same phenotypic

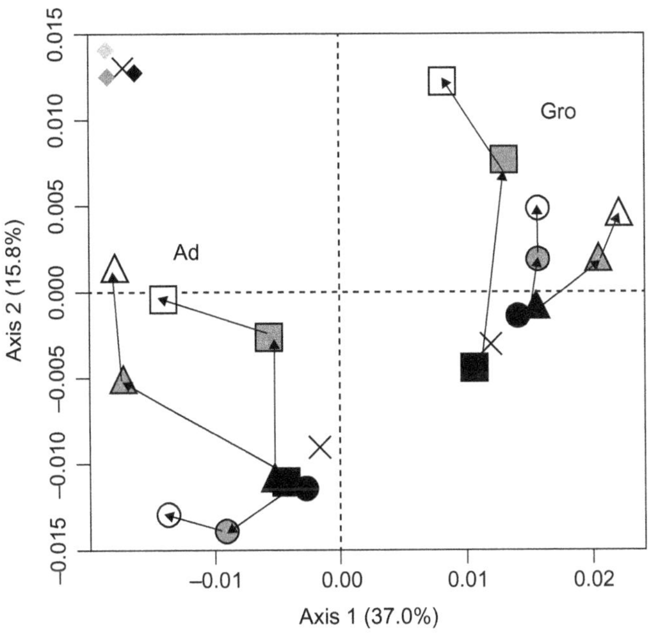

Figure 7.2: Genome-wide evolution during the first 50 generations in the laboratory. This principal coordinate analysis compares F_{ST} values across the populations and generations. In contrast with the rapid phenotypic convergence (Figure 7.1), the populations maintained high genetic differentiation over time. Ad: Adraga founders; Gro: Groningen founders. Black diamonds: long-term (control) populations. Arrows connect generations of the same replicate population; X symbols show values for the founders (adapted from Figure 3A in Seabra et al. [2018]).

outcomes could be achieved through different genetic paths, an important issue in the literature (Orgogozo, 2015; Tenaillon et al., 2012). Moreover, detailed analysis of the frequency changes in selected alleles across generations 6, 25, and 50 revealed a panoply of patterns. There was not a simple dichotomy between steady linear changes, on the one hand, and fast initial changes with later plateauing, on the other hand, in contrast to the study by Orozco-TerWengel et al. (2012). However, the different statistical approaches used in these studies may have led to the disparate conclusions.

7.2.7. *Follow-up studies on the predictability of laboratory adaptation*

Given the importance of understanding the repeatability of evolutionary dynamics, this study was repeated three years later using new samples of founders from the same locations. The results confirmed the generally rapid phenotypic convergence across all populations from different locations and different years (Simões et al., 2017). Ongoing genomic analysis suggests that evolution is less repeatable at the genetic level (in progress). This work shows the importance of analyzing many populations at different levels of biological organization and over many generations before making broad generalizations.

7.3. The Ives *D. melanogaster* Lineage and Descendants

7.3.1. *Origins and motivations*

Charlesworth spent the 1970s developing useful population genetics theory for age-structured populations (Arnold & Rose, 2023; Charlesworth, 1980). Of greatest practical significance was his application of this theory to the evolution of aging, which motivated him to test his findings using *Drosophila*.

The *Drosophila* stock that he chose came from Philip Ives, who had been studying a long-established *D. melanogaster* population that was endemic to apple orchards in Massachusetts (Ives, 1970). This particular choice would prove to be significant for later research on diet, as will be discussed below. Charlesworth took care to avoid inbreeding of the descendants of the founding sample of 200 gravid females, which was important for all subsequent work with the Ives lineage. It is customary in many *Drosophila* laboratories to maintain "outbred" populations using just two vials or bottles: one containing older medium and flies, and the other started with eggs laid by females from the "old" vials, these females then being discarded. Once the "new" vial has visible adults, it becomes the "old" vial, and the previous "old" vial is discarded. Such "wild-type" populations have effective population sizes of less than 100. Instead of using this culture regime, the Ives-sampled population was maintained using either 16 pint-sized bottles containing maize-molasses medium (before 1981) or 20-40 8-dram vials containing banana-syrup medium (after 1981). All adults were mixed together at 14 days of age, at 23–25°C after using CO_2 as anesthesia. Then these adults were placed in the same number of culture vessels with fresh medium. Once they had laid at least 1,000 eggs, the adults were discarded, starting a new discrete generation. Effectively, Charlesworth established a lab population with an effective population size approaching 1,000 and discrete generations (see Mueller et al., 2013). This basic culture regime has been maintained to this day, with long-sustained, genome-wide genetic variation characterizing this "IV" population and its derivatives, as Charlesworth intended (see Graves et al., 2017; Phillips et al., 2016).

Charlesworth's goal in maintaining a less-inbred IV *Drosophila* population was to test his favored evolutionary genetic mechanism for aging: *mutation accumulation*. In evolutionary genetics, the term "mutation accumulation" refers to the accumulation of deleterious mutations that affect fitness and its components. An early theoretical finding of Charlesworth and Williamson (1975) was that mutation accumulation was predicted to produce more genetic variation for age-specific life-history components with increasing age. [This prediction would later be qualified in terms of the distribution of phenotypic effects among such deleterious mutations as a function of age (Charlesworth & Hughes, 1996).] To test this prediction, Charlesworth persuaded Michael Rose to use sib analysis to estimate the genetic variances of age-specific fecundity in the Ives population during the 1970s. Unfortunately for Charlesworth's prediction, no such increase in genetic variances was detected (Rose & Charlesworth, 1980, 1981a). Instead, there was a strong negative genetic correlation between early fecundity and longevity (Rose & Charlesworth, 1981a), providing support for an alternative evolutionary genetic mechanism for the evolution of aging, *antagonistic pleiotropy* between early reproduction and later survival (Rose, 1985; see also Arnold & Rose, 2023, for a discussion of the history of this hypothesis).

While performing this large sib analysis, Rose happened to read the literature on the "Lansing effect," which referred to the supposed decline in longevity of populations maintained at later ages (Arnold & Rose, 2023). This effect was absent in homozygous *Drosophila* stocks (Comfort, 1953) or reversed in less inbred stocks (Wattiaux, 1968a, 1968b). Rose interpreted the latter finding as suggestive of the potential for laboratory evolution to produce postponed aging in a relatively short period of time (see Rose, 2005). He then proceeded to culture a derivative of the Ives *Drosophila* stock that reproduced at later ages for 12 generations and then compared its longevity with that of its ancestral stock. The results were that these late-reproduced flies had increased longevity but reduced early fecundity, again supporting the importance of antagonistic pleiotropy in the evolution of aging (Rose & Charlesworth, 1980, 1981b).

For Rose, the more important point was that it was far easier to experimentally evolve *Drosophila* populations than to study their quantitative genetics using methods like sib analysis. Accordingly, from 1980 onward, his chief focus was on the use of experimental evolution to test a wide range of hypotheses about the evolution of life history and other functional traits in derivatives of the Ives population.

7.3.2. *Evolution of aging in B and O populations*

An obvious problem with the single late-reproduced population created by Rose in the 1970s was that it was unreplicated. Thus, its key findings could have arisen from a coincidental concatenation of drift and selection, rather than a systematic effect of a change in the selection regime. Accordingly, in 1980 Rose established 10 derivatives of the IV population, each starting with eggs laid by a collection of about a thousand adults from a single IV generation. Five of these derivatives were maintained using the IV culture regime and were called the B1 to B5 populations. Five other populations were maintained using eggs laid by females at progressively greater ages, topping out at 70 days of age by 1981, the O1 to O5 populations. Starting in 1982, and continuing ever since, these two groups of five populations have been compared for a wide range of traits (Rose et al., 2004).

The first assays performed on the B and O populations focused on life-history traits, from viability to longevity to age-specific fecundity (Rose, 1984b). It is a notable feature of these assays that adults were handled in vials or cages at moderate densities with both sexes present. Most experiments in gerontology are conducted using single-sex cohorts, in which mating, and thus reproduction, is precluded. This approach creates a generic problem of *genotype-by-environment interaction* ("GxE"), which tends to bias genetic correlations toward positive values, either in direct quantitative genetic assays (e.g., Service & Rose, 1985) or in assays of the effects of experimental evolution (e.g., Clare & Luckinbill, 1985; Service et al., 1988). Thus, from the beginning, the Rose laboratory was concerned with avoiding GxE artifacts. Despite this, GxE artifacts have nonetheless arisen in the life-history assays of the Ives lineage, artifacts that take years to sort out (e.g., Leroi, Chen & Rose, 1994; Leroi, Chippindale & Rose, 1994).

One of the most serious artifacts impinging on studies of life-history traits was, however, consistently avoided: *inbreeding depression*. By reducing additive genetic variation, inbreeding can produce failures of response to experimental evolution (e.g., Comfort, 1953). It can also produce artifactual increases in genetic correlations (e.g., Rose, 1984a). Inbreeding depression and GxE together undermine much of the literature in gerontology (Arnold & Rose, 2023) and some of the experimental literature on life-history evolution more generally. Both of these problems were avoided or mitigated in the Rose laboratory's studies of experimental evolution in the Ives lineage of *Drosophila* populations.

With respect to the evolution of life history, the chief result from decades of assays of the B and O populations is that aging is readily postponed by delaying the age of first reproduction in experimentally evolved populations over dozens to hundreds of generations (Burke et al., 2016; Rose et al., 2004; recall Figure 2.2). This postponement is accompanied by reduced early reproduction, provided that the reproductive assay conditions are similar to those of the matched B populations that serve as the controls (Leroi, Chippindale & Rose, 1994). These findings support the basic theory that the evolution of aging reflects the declining force of natural selection with age (Hamilton, 1966). They also support the importance of antagonistic pleiotropy as an evolutionary genetic mechanism, both in maintaining genetic variation and in producing a trade-off between early reproduction and longevity (Arnold & Rose, 2023).

7.3.3. *Evolutionary physiology of aging and function*

In 1982, a technician in the Rose lab inadvertently performed a perfectly controlled experiment to compare the response to acute stress on B and O cohorts at the same adult age by denying food to a set of young adults of each type overnight. Even at fairly early adult ages, the O flies survived the starvation stress in large measure, while the B flies largely died off. This set the stage for decades of experimentation on the evolutionary physiology of populations derived from the IV lineage.

Given this starting point, research on evolutionary physiology in the B and O populations focused on comparing the two sets of populations with respect to stress resistance. The work of Phil Service established clear differences between B and O populations across adult ages for resistance to total starvation, ambient ethanol, and acute desiccation, but not to humidity-controlled heating (Service, 1987; Service, 1989; Service et al., 1985). The stress-resistance assays all proceeded under conditions that lacked food or mating opportunities, though the flies were not denied the opportunity to mate prior to the stresses. This pioneering work set the stage for two branches of subsequent work on the physiology of experimentally evolved *Drosophila*: selection on stress-resistance traits themselves and mechanistic work on the physiological foundations of stress resistance.

Thus, starting in the 1980s, the Rose lab began to select on B- and O-derived populations for acute stress resistance. For example, Rose et al. (1990, 1992) selected for desiccation resistance in derivatives of the O populations, as well as selecting for starvation resistance in derivatives of both B and O populations. In each case, selection involved passing adults through phases of acute stress in cages with the same basic stress regimes as those used in the assays of Service (1987). The survivors of the acute stress, most of whom were females, were then allowed to reproduce. Levels of stress resistance increased substantially in the "S" (starvation) populations, for example, sometimes approaching 10-fold. These same protocols have since been used in additional populations of the Ives lineage (e.g., Archer et al., 2007, 2003; Kezos et al., 2017), with similarly strong responses to selection.

Adding to the growing spectrum of Ives-derived populations were ones selected for faster development and reproduction. These were called the "A" populations (e.g., Chippindale et al., 1997), some derived from B ancestors and others from O ancestors. Joining them were reverse-selected or "relaxed selection" populations, some derived from O populations and others from stress-selected populations (e.g., Passananti, Beckman & Rose, 2004; Passananti, Deckert-Cruz et al., 2004; Service et al., 1988; Teotónio & Rose, 2000). In keeping with the expectation of widespread antagonistic pleiotropy, it was usually the case that reverse selection resulted in the loss of the increases in functional traits that had been achieved during "forward selection," although not as quickly for some traits as for others (e.g., Burke et al., 2016; Phillips et al., 2018; Teotónio & Rose, 2000).

Thus, from 1990 to 2020, the Rose laboratory had a wide range of populations altered by experimental evolution, all derived directly or indirectly from the IV population. At times, these varied populations numbered in the hundreds, but 60–100 populations were typical, all available for physiological research. In terms of aging, it was generally found that moderate increases in stress resistance by selection were associated with increased longevity (e.g., Rose et al., 1992). However, selection for very high levels of stress resistance was associated with reduced longevity (e.g., Archer et al., 2007; Phelan et al., 2003) and other impairments, such as reduced heart robustness (Kezos et al., 2019). The physiological mechanisms that underlie stress resistance were found to be heterogeneous. The experimental evolution of starvation resistance depended entirely on caloric reserves, chiefly fat, but not on metabolic rate (Djawdan et al., 1998). Desiccation resistance depended on water reserves and water loss rates, but not on tolerating low water content (Gibbs et al., 1997; Nghiem et al., 2000). Interestingly, the experimental evolution of water reserves and water loss rates was shown to proceed independently of each other by

Archer et al. (2007), who monitored their trajectories in response to desiccation selection.

With respect to the diversity of cell-molecular mechanisms of aging widely promulgated in the gerontological literature, none were found to be important physiological determinants or correlates of the experimental evolution of aging. Instead, *Drosophila* aging and related phenotypes appear to evolve as a function of organismal and organ-level physiology, such as fat content, much as mortality and morbidity in humans correlate with organismal and organ-level functions. This pattern would continue in genomic research with the Ives lineage, discussed below.

7.3.4. *Experimental evolution of late life*

In 1992, two labs published results suggesting that large cohorts of two Dipteran (fly) species stopped aging at late adult ages (Carey et al., 1992; Curtsinger et al., 1992). Furthermore, earlier research by Greenwood and Irwin (1939) was rediscovered, their demographic data indicating a similar cessation of aging among humans over 90 years of age, at least among Europeans who ate a pre-industrial or "organic" diet during the 19th and early 20th centuries. These findings were offered as a refutation of virtually all gerontological and evolutionary theories of aging, which had all presumed that aging proceeds without remit until every member of a cohort died.

The cessation of aging has now been observed for age-specific survival, fecundity, and virility across a wide variety of species whenever sufficiently large cohorts are used (reviewed in Mueller et al., 2011). This research has thus refuted the core mainstream gerontological assumption that aging is some kind of cumulative process of deterioration (Arnold & Rose, 2023). What was not clear at first was whether the evolutionary theory of aging as developed by Hamilton (1966) and Charlesworth (1980) would survive.

Mathematical analyses, however, showed that the eventual cessation of aging corroborates, rather than refutes, the evolutionary analysis of aging (Charlesworth, 2001; Mueller et al., 2011; Mueller & Rose, 1996). Theoreticians had simply failed to look at the implications of Hamilton's forces of natural selection for the likelihood of aging continuing without remit.

As with the way that the Hamiltonian forces of natural selection impact the start of aging, it turned out that similarly simple and manipulable factors controlled the timing of the cessation of aging: namely, the last ages of survival and reproduction that prevailed during the evolutionary history of a population (Mueller et al., 2011). These parameters could be readily altered in the *Drosophila* populations of the Ives lineage and indeed already had been in populations prior to 1992. It turned out that experimentally changing these late-in-life parameters led to rapid evolution that altered the timing of the cessation of aging across the life-history traits of mortality, fecundity, and virility (Rauser et al., 2006; Rose et al., 2002; Shahrestani et al., 2012). Altogether, research on late life provided another case for the value of experimental evolution to rigorously test evolutionary theories.

An additional payoff from this work was an increased focus on the use of very large cohorts in studies of the experimental evolution of life-history traits. A striking benefit of this shift was the use of experimental evolution to test the idea of Hamilton's forces of natural selection scaling the speed of age-specific adaptation to sustained shifts in qualitative features of a population's environment that impact ages of reproduction and lifespans, an important issue in the application of evolutionary theory to human aging (Mueller et al., 2011). Specifically, the broadened theory required to account for late life implies that age-specific adaptation to an environmental change will be faster at juvenile and early adult ages compared to later adult ages. This was tested in the Ives radiation using a comparison of early reproducing A populations with later-reproducing C populations on three kinds of diets: an entirely novel diet (orange-based), the diet sustained during lab domestication (banana-syrup), and a mushed-apple diet like the one that supported the ancestral wild population of Ives (Rutledge et al., 2020). The results of this study were striking: in the A populations, a crude emulation of the long-abandoned apple diet was best at later ages, while in the C populations, which sustained adaptation to the lab diet at later ages, the apple diet was inferior for a longer period of adult life. This work was yet another illustration of the power of experimental evolution to test evolutionary theory (Arnold & Rose, 2023).

7.3.5. *Omics studies of the Ives populations*

Genomics research has great power compared to studies of candidate genes for understanding experimentally evolved populations. While early studies of candidate genes in Ives-derived populations yielded promising results (e.g., Deckert-Cruz et al., 1997), the question remained as to whether the specific loci studied in that research were representative of what was occurring across the entire genome. An early unbiased estimate of genome-wide change was provided by a two-dimensional gel electrophoresis study comparing B and O populations, which implicated a few percent of all loci (Fleming et al., 1993). The *D. melanogaster* genome has about 14,000 transcribed genes, as well as thousands of sites with minor structural variation and transposable element insertions. Across about 20,000 sites, then, differences between the B and O populations at 4% of them imply changes at about 800 sites.

At first, genome-wide research on the Ives populations failed to use the full power of the system, with replicate populations in a given treatment being pooled prior to shotgun sequencing (Burke et al., 2010). That pooling meant it was impossible to tell if a given allele's frequency had increased moderately in all of the replicates or, alternatively, it had increased more dramatically but in only one replicate. Nonetheless, that early research again suggested that many sites across the fly genome were involved in the experimental evolution of the Ives-derived populations. Properly designed and executed genome-wide sequencing of 30 Ives-derived populations yielded estimates of hundreds of genomic sites with statistically detectable responses to experimental evolution (Graves et al., 2017), in keeping with the crude early estimates of Fleming et al. (1993).

Genome-wide transcriptomic research comparing A and C cohorts at day 21, an age at which A cohorts are aging demographically but C cohorts are not, again found hundreds of transcripts differentiating these two types of population as a result of their experimental evolution (Barter et al., 2019). Recall, however, that the number of transcriptomic differences may be greater than the number of genomic differences, since a difference at one locus can affect the expression of many genes (see Chapter 4). Though the number of metabolites analyzed was fairly small, metabolomic comparisons among various Ives-derived populations again suggested that the levels of many were altered in response to differences in their experimental evolutionary regime (Phillips, Arnold et al., 2022).

Arnold (2025) recently conducted something of a capstone project for the Ives-derived adaptive radiation. He started from the evidence for convergence among Ives populations undergoing shared recent selection regimes, despite contrasting evolutionary histories (e.g., Burke et al., 2016; Graves et al., 2017). This led to the hypothesis that specific selection regimes imposed on the Ives populations had genome-wide site-by-site attractor states, to which populations would converge. He then performed a double reverse-evolution experiment: 10 populations that had converged under the A-type early reproduction regime were switched to the later-reproduction C-type selection; at the same time, 10 C-type populations were switched to the early-reproduction A-type regime. As expected, the life-history traits largely converged to those that are characteristic of the newly imposed regimes.

The genomic findings were more striking. Although Arnold (2025) only analyzed SNP frequencies, the scope of the data was exceptional by the standards of *Drosophila* experimental evolution, encompassing a total of 40 populations with multiple generations sampled and sequenced. In Figure 7.3, the antiparallel dynamics for a single SNP show a pattern of switching from one stable state to another, consistently among replicates. Again, this is in keeping with the expectations inferred from Graves et al. (2017), but in the Arnold study, the trajectories transparently illustrate the hypothesized switching in action.

At the level of genome-wide patterns, the results are perhaps more surprising. Figure 7.4 shows the dynamics of genome-wide average heterozygosity under the antiparallel selection regimes. These genome-wide dynamics evidently emulate the pattern at the level of individual SNPs, with a global transformation of the genomics of the populations undergoing selection, rapidly switching from one heterozygosity level to another.

The analysis performed by Arnold (2025) went even further, slicing up the genome into 50-kb segments and then assessing whether the individual segments underwent detectable frequency changes under the double-reverse selection regimes. With the genome parsed in this manner, about 44% of the genome included SNPs that had statistically detectable changes in frequency. There is extensive linkage disequilibrium in the

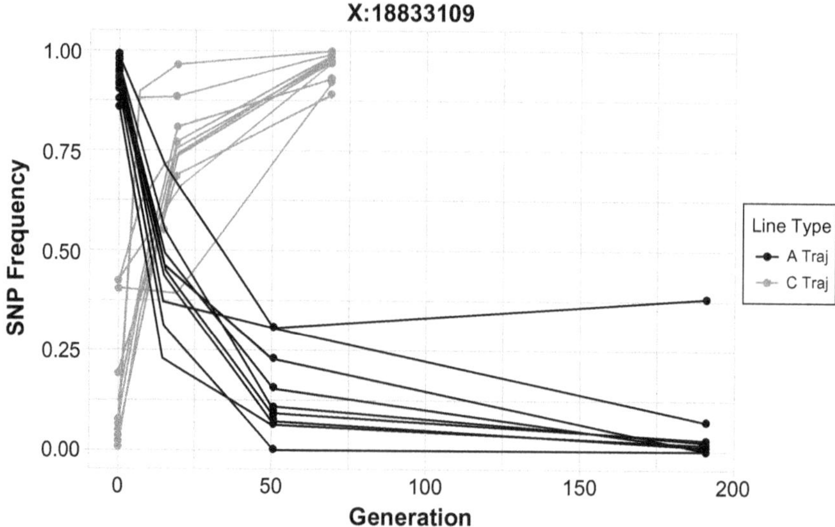

Figure 7.3: Trajectory of a highly significant SNP across generations. Allele-frequency trajectories are shown for a single SNP located on the X chromosome (position 18,833,109), which exhibits a strong response to selection after Bonferroni correction ($p = 2.24 \times 10^{-13}$). Each line represents one population, illustrating allele-frequency dynamics in the newly A-selected treatment and the newly C-selected treatment lineages over ~185 and ~65 generations, respectively. Figure adapted from data presented in Arnold (2025).

50-kb range, but even so, this finding implies that a large fraction of the genome includes variants that are affected by selection, either directly or indirectly through linkage (i.e., genetic draft). As a point of comparison, Linder et al. (2022) found that allele frequencies in most of the genome were affected, again either directly or indirectly through linkage, by at least one of six selection regimes in large, highly polymorphic, and repeatedly outcrossed populations of yeast. Put another way, selection can wield extensive sway across the genome, transforming it widely when selection is sufficiently strong and sustained.

The overall "omics" picture of the Ives-derived adaptive radiation is one of great complexity. There are hundreds of sites in the genome that have undergone moderate shifts in the frequencies of allelic variants, most of which have not gone to fixation. Furthermore, these shifts are largely reversible when experimental evolution proceeds in reverse

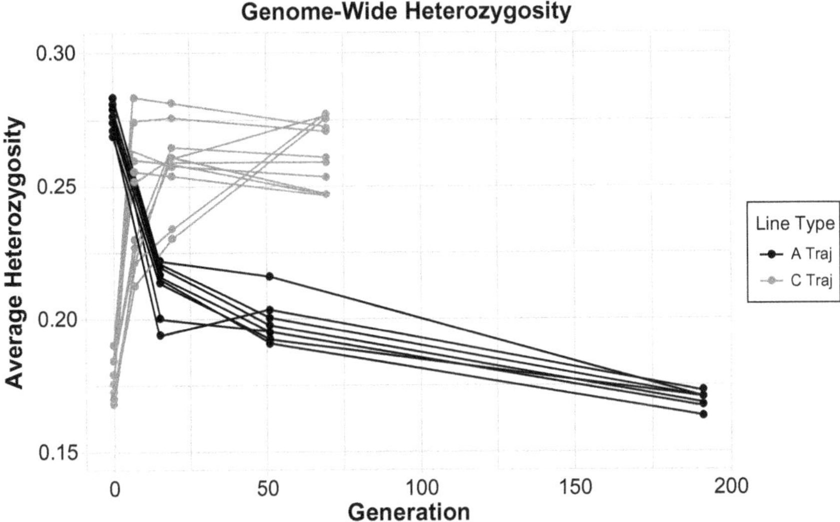

Figure 7.4: Genome-wide average heterozygosity across major chromosome arms: Heterozygosity was calculated from SNP data across all major *Drosophila melanogaster* chromosome arms. Values begin with the ancestral populations and then follow the newly A-selected trajectories (C→A) and newly C-selected trajectories (A→C) lines across successive generations, each line representing a single population. Figure adapted from data presented in Arnold (2025).

(Arnold, 2025; Graves et al., 2017), just as phenotypic shifts are also largely reversible (e.g., Arnold, 2025; Burke et al., 2016).

Underlying these results, a broader evolutionary genetic pattern emerges. On the scale of hundreds of generations, many sites in the genome harbor functionally significant variation. That variation appears to be maintained by some type of balancing selection, with antagonistic pleiotropy that produces tradeoffs playing a substantial role in that balancing selection. We have referred to this pattern of Mendelian evolutionary genetics as "neo-balance" (Arnold, 2025; Arnold & Rose, 2023), an extension of the "balance" perspective that predates the advent of molecular genetics (see Lewontin, 1974). The neo-balance perspective can be contrasted with the "neo-classical" perspective, which emphasizes the predominance of neutral and deleterious genetic variation while downplaying the importance of standing genetic variation of functional value. The neo-balance view allows that much genetic variation is merely neutral or mildly deleterious, while nonetheless emphasizing that the genetic foundations of adaptation in Mendelian populations are often found in selectively main-tained genetic polymorphisms that can respond quickly to changes in selection regime, such as those imposed during experimental evolution.

7.4. Other Noteworthy Experiments with Sexual Organisms

7.4.1. *Temperature adaptation in hybridized* D. simulans

Barghi et al. (2019) conducted an impressive and revealing evolution experiment with *D. simulans*. They started by hybridizing 202 fairly inbred isofemale lines, with the resulting progeny used as founders to cre-ate 10 highly polymorphic starting populations. Over the course of 60 generations, these populations were subjected to a novel culture regime of alternating 12 h dark and 12 h light at temperatures of 18°C and 28°C, respectively. They then monitored phenotypic and genomic evolution using samples taken every 10 generations.

The 10 populations showed strongly parallel phenotypic responses, especially in lipid content and metabolic rate. They also found frequency changes in a number of SNPs in genes involved in lipid metabolism and other aspects of catabolism. However, the specific genetic changes were not uniformly consistent across the 10 experimental populations. More commonly, parallel SNP frequency changes were shared by four to six replicate populations. Extensive numerical analysis showed that the SNP frequency changes did not conform to patterns expected with conventional selective sweeps (see Burke, 2012; Burke et al., 2010).

The chief disparity relative to results obtained from the experimental Ives-derived populations (e.g., Graves et al., 2017) was the paucity of parallelism at the genomic level. Interestingly, parallelism increased as the omics analyses proceeded from genomic to transcriptomic to metabo-lomic levels (Lai et al., 2023), approaching the striking convergence shown by phenotypes at the level of organismal physiology. Schlötterer (2023) has argued that the differences between his lab's results and those obtained for the Ives-derived lineages reflect differences in the founding populations used to initiate the evolution experiments. Support for this view is found in another elegant study from his laboratory: Burny et al. (2021) found much greater convergence during thermal adaptation among replicate populations started from founders obtained by crossing just two of the 202 isofemale lines that were used to start the thermal adaptation experiment of Barghi et al. (2019).

The population used to found the many experimental populations derived from the Ives lineage had been kept in the lab for more than 100 generations before the original B and O populations were created (Rose, 1984b). The Ives-derived B populations have long sustained abundant genome-wide variation (Phillips et al., 2016), but they are not the products of hybridization, nor were they subjected to sustained inbreeding in the lab. Random founder effects almost certainly occurred during the establishment of the Ives stock in the lab, as suggested by the microsatellite findings of Santos et al. (2012, 2013) with *D. subobscura,* discussed earlier in this chapter. But these random founder effects would be shared among all the descendants of the Ives stock. By contrast, the founding population of the Barghi et al. (2019) study necessarily had extensive linkage disequilibrium that would have interacted with selection in the 10 descendant populations undergoing thermal adaptation, with each of the 10 descendant populations subject to an historically unique interaction of linked alleles and selection. Such historicity is a basic feature of the nonlinear complexities of multi-locus selection systems (e.g., Nagylaki, 1977).

It is also worth noting that an evolution experiment with yeast that used a founding population derived by crossing four homozygous stocks yielded parallel genomic changes (Burke et al. 2014), a result in keeping with the parallelism observed by Burny et al. (2021). Overall, the pattern of these results supports Schlötterer's (2023) contention that the nature of the founding populations in evolution experiments can strongly influence the observed genomic outcomes.

7.4.2. *Selection for wheel-running behavior in mice*

Ted Garland's laboratory has performed one of the best-designed, longest, and most thoroughly analyzed selection projects involving mammals, and perhaps any vertebrate, outside of agricultural breeding programs. The experiment was started in 1993 with a base population consisting of 112 males and 112 females of *Mus musculus* using the relatively outbred "ICR" albino strain, which in turn had origins dating back to 1947. Over most of the time between the strain founding and the start of Garland's experiment, the ICR stocks were maintained using about 1,000 breeding females (Swallow et al., 1998), thereby sustaining a reasonable level of genetic variation.

Eight lines were derived from the initial sample of 224 mice: four unselected controls and four lines selected for high voluntary wheel-running, called "High Runner" (HR) lines. Selection was conducted within families, with 2–3 individuals chosen to reproduce from each of 10 to 13 families, and sib mating was precluded. Thus, these lines had quite low effective population sizes and, consequently, estimates of inbreeding have steadily increased during the experiment (e.g., Careau et al., 2013).

The HR lines showed a steady increase in average wheel-running times over the course of about 20 generations, after which they appear to have plateaued (Careau et al., 2013). The genomes of the eight lines have been analyzed extensively (e.g., Hillis et al., 2020). However, only a few SNPs are significantly differentiated between the HR and control lines after correcting for multiple tests across the entire genome. Interestingly, there has been a pronounced increase in the frequency of a "mini-muscle" allele in two of the four HR lines (Hillis & Garland, 2023). In the homozygous state, that allele is associated with running faster but for a shorter duration (Hillis & Garland, 2023).

Of greater importance is the extensive physiological characterization of the HR and control lines, with more than 150 studies to date. Many of these studies have documented parallel evolution of differences between the two sets of lines (see Swallow et al., 2009, for a review of the early physiological work on these lines). The detail and insights from this body of work define a peak for the field of evolutionary physiology, with findings that range from the characterization of muscle development and structure to numerous differences in neurological function. This progress reflects not only the careful design of the Garland experiment but also the ability of mammalian evolutionary physiologists to draw on the extensive tools and ideas of medical physiology. As a consequence, it provides a new standard that other studies in evolutionary physiology could emulate.

7.5. Chapter Summary

Experimental evolution with obligately sexual and multicellular organisms is in many respects far more challenging, as a practical matter, than experimental evolution with asexual microorganisms. Culturing these larger organisms requires greater space per individual, a difficulty compounded because large population sizes are required to avoid inbreeding and generate reliable inferences. Nonetheless, species in the genus *Drosophila* have provided material for many productive evolution experiments, thanks to their small body sizes, their relative resistance to culture infection, and their long-standing use as research workhorses across a wide spectrum of biological disciplines. We have illustrated these achievements chiefly through discussion of the research from the laboratories of Matos, Rose, and Schlötterer. Of course, other *Drosophila* labs have also conducted evolution experiments, but this primer cannot provide an exhaustive survey.

A chief point of difference among evolution experiments with sexual species is the nature of the founding populations used to start specific projects. When such founding populations are complex assemblages of many isofemale lines, as in Barghi et al. (2019), parallel evolution at the genomic level has proven to be rare, at least at moderate population sizes. [Using sexual, outcrossed yeast and much larger populations, Linder et al. (2022) found more genomic parallelism.] But when there is less complexity in the founding population, experimental evolution in sexual populations can feature extensive genomic parallelism and convergence, as shown by Graves et al. (2017) and Burny et al. (2021).

Whatever the degree of genomic complexity, these evolution experiments with sexually reproducing organisms have demonstrated a great deal of uniformity among replicates at the levels of physiology and life history, with the conspicuous exception of a large-effect "mini-muscle" mutation arising in two of the HR mouse selection lines (Hillis & Garland, 2023). That exception occurred in the evolution experiment with the lowest effective population size of the studies considered in this chapter.

We do not expect to see such unusual events in evolution experiments that feature larger effective population sizes, leaving aside studies where inbreeding is deliberately introduced as part of the experimental design. Perhaps more exceptions might occur, even in large populations of sexual organisms, if an evolution experiment could be run for so long that new mutations, rather than the extensive variation present in the founders, eventually became the main source of beneficial alleles.

We defer such questions of larger scientific significance to the final chapter of this book, because we regard the contrasts between experimental evolution with and without sex as highly informative.

Chapter 8

Conclusions

8.1. State of the Art Summarized

8.1.1. *Two kinds of systems for experimental evolution*

In some respects, the field of experimental evolution can be divided into two rather different scientific worlds. On the one hand, there are those experimental systems in which the organisms, generally microbes, do not undergo sexual recombination. In that case, much of the analysis focuses on understanding mutation and selective sweeps, though one must also account for random genetic drift as well as the possibility of frequency-dependent selection maintaining genetically distinct lineages (e.g., Good et al., 2017; Rainey & Travisano, 1998; Rozen & Lenski, 2000). In any case, in these asexual systems, there is genome-wide linkage disequilibrium, and selection acts on the genetic variance in overall fitness.

On the other hand, there are evolution experiments performed with outbreeding sexual populations. In these experiments, new mutations may contribute a bit to the genetic variation. But of far greater importance are the initial genetic variation, patterns of recombination, linkage, dominance, and epistasis. In these systems, linkage disequilibrium is local, and responses to selection depend strongly on additive genetic variances and covariances.

Despite these important differences, many of the key strategies of experimental design, genomic analysis, and functional interpretation are general across all systems used in experimental evolution. Among these are the importance of replication and controls, the nuances of assaying fitness or its components, the value of genome-wide sequencing, and the necessity of collecting data across many generations. With respect to interpretation, all these systems also share the challenges of parsing the causal hierarchy from genome to transcriptome and metabolome to morphology, physiology, and behavior, and ultimately to the Darwinian fitness of the organisms.

8.1.2. *The importance of founders*

One of the most important, but easily overlooked, aspects of designing an evolution experiment is choosing appropriate founders with which to start the populations. It was important that Lenski started the Long-Term Evolution Experiment (LTEE) with a strain of *Escherichia coli* that lacked the potential for horizontal gene transfer (Lenski, 2023; Lenski et al., 1991). This choice ensured genome-wide linkage, which allowed a readily scored genetic marker to be used as a proxy for ancestral and derived bacteria in competition assays to measure Darwinian fitness. Thanks to a choice made by Charlesworth, the Ives lineage of the *Drosophila melanogaster* system developed by the Rose lab (e.g., Rose et al., 2004) began with an ancestral population that had been cultured at a reasonable population size in the laboratory for more than 100 generations. Thus, many issues that bedeviled later work using recently hybridized founders (e.g., Barghi et al., 2019), such as genome-level irreproducibility, did not affect research with the Ives-derived lineages (e.g., Barter et al., 2019; Graves et al., 2017).

8.1.3. *Parallelism, peaks, and plateaus*

Questions and data bearing on patterns of evolutionary parallelism (or convergence when starting with different ancestors) and fitness peaks or plateaus have been of great interest for the field of experimental evolution. In the LTEE, the system with the most generations, there has been a long, slow deceleration in the trajectory of mean fitness (Lenski et al., 2015; Wiser et al., 2013). Nonetheless, mean fitness continues to increase, as ongoing mutations continue to discover small beneficial effects. Moreover, there have been striking examples of parallelism at the genetic (e.g., Woods et al., 2006), transcriptomic (e.g., Cooper et al., 2003), morphological (e.g., Grant et al., 2021), and fitness (e.g., Lenski et al., 2015) levels. With regard to parallelism, the large population sizes, which number in the millions even during the transfer bottlenecks, are undoubtedly important. On the other hand, there is also clear and compelling evidence for multiple genetic pathways for adaptation in both the LTEE itself (e.g., Blount et al., 2008, 2012; Couce et al., 2024) and other experiments that derive from the LTEE (e.g., Tenaillon et al., 2012).

In experiments with sexual populations, including those derived from the Ives *D. melanogaster* lineage, there is a clear parallelism among the replicate populations that have faced the same selective regime for at least 100 generations at the levels of life history (e.g., Burke et al., 2016), genome (Graves et al., 2017), transcriptome (Barter et al., 2019), metabolome (Phillips, Arnold et al., 2022), and physiology (e.g., Kezos et al., 2023). Similarly, striking phenotypic convergence has been observed in the *D. subobscura* system used to study evolutionary domestication (Fragata, Simões et al., 2014; Simões et al., 2017).

8.1.4. *Evolutionary physiology and omics*

One of the most sustained areas of progress in experimental evolution has been the increasing detail and depth of mechanistic research between the levels of genome and life history. This research started with fairly crude analyses of whole-organism physiology (e.g., Graves et al., 1988; Service et al., 1985). But in experiments with *Drosophila*, it has since progressed to detailed resolution of the biochemistry underlying such physiological functions, from aggregate biochemistry (e.g., Djawdan et al., 1998) to specific metabolites (e.g., Lai et al., 2023; Phillips, Arnold et al., 2022) and transcripts (e.g., Barter et al., 2019; Remolina et al., 2012). In experiments using bacteria, progress has similarly provided great detail in the resolution of such features as aggregate biochemistry (e.g., Peng et al., 2018; Turner et al., 2017), specific metabolites (e.g., Blount et al., 2020; Großkopf et al., 2016; Quandt et al., 2015), and transcriptomes (e.g., Cooper et al., 2008, 2003; Favate et al., 2022).

8.1.5. *Genomics of experimentally evolved populations*

In the early days of genetic research on the products of experimental evolution, analyses focused on specific candidate genes (e.g., Cooper et al., 2003; Deckert-Cruz et al., 1997), but whole-genome sequencing was clearly the technology most appropriate for the field. Barrick et al. (2009) and Burke et al. (2010) pioneered this approach for experimentally evolved *E. coli* and *D. melanogaster*, respectively. Subsequent genome-wide research with the products of experimental evolution has become even more powerful and revealing (e.g., Good et al., 2017; Graves et al., 2017; Tenaillon et al., 2016, 2012). Indeed, we believe that the power of this "Evolve & Resequence" paradigm (Schlötterer, 2015) is much greater than that of the far more widely employed Genome-Wide Association Studies, which often suffer from an inability to associate particular genomic sites with important phenotypes (e.g., Ivanov et al., 2015).

8.2. Applications to Big Questions in Biology

8.2.1. *Adaptation*

In the past, we have referred to experimental evolution as a "Wonderland" for the study of adaptation (Rose et al., 1996). Thanks to the control of environmental conditions, replication of populations, and the ability to follow trajectories of evolutionary change, there are few systems for the study of Darwinian adaptation that are more powerful (see Rose & Lauder, 1996).

With respect to the pattern of very long-term adaptation in asexual populations, Lenski's LTEE has defined a standard of excellence that few other laboratories have approached. While the evolution of life-history components of fitness, and thus adaptation, has been extensively studied using *Drosophila* in the Rose and Matos laboratories, a definitive analysis of the evolution of fitness itself in sexual species is very difficult, as we discussed in Chapter 4. In particular, the study of sexual components of fitness presents extreme challenges for observation and measurement.

8.2.2. *Aging*

The phenomenon of aging was one of the first to be studied using the tools of experimental evolution, both inadvertently (Wattiaux, 1968a, 1968b) and deliberately (Rose & Charlesworth, 1980; Sokal, 1970). Experimental evolution in the Ives-derived *D. melanogaster* populations (e.g., Rose et al., 2004) as well as other *D. melanogaster* lineages (e.g., Luckinbill et al., 1984) has allowed substantive progress on questions concerning the forces that control aging and its underlying evolutionary genetic mechanisms. Of note is the recent turn to the issue of the possible cessation of aging, a topic that received negligible attention before 1990 (see Mueller et al., 2011). Experimental evolution has played a leading role in establishing the evolutionary basis of aging as one of the more successful applications of evolutionary theory in recent times (Arnold & Rose, 2023; Rauser et al., 2009).

8.2.3. *Sex*

There are three discernible evolutionary issues that arise with sex in evolutionary biology: the contrasts between the evolution of asexual and sexual systems, the evolutionary origins of sex, and the maintenance of sex, especially in its anisogamous form (i.e., where the gametes that fuse to produce offspring are unequal in size).

The first of these three issues has benefited greatly from the comparison of asexual and sexual systems used in experimental evolution. In strictly asexual populations, mutation and selective sweeps are conspicuous and important features of the evolutionary process, with genome-wide variation repeatedly expunged from populations (e.g., Blundell et al., 2019; Tenaillon et al., 2012) or, at least, from subpopulations (Good et al., 2017). As such, adaptation in asexual organisms requires sufficiently large population sizes to provide the mutations on which selection can act. By contrast, sexual populations can maintain abundant genetic variation for long periods of evolutionary time, even at relatively small population sizes (see Graves et al., 2017; Phillips et al., 2016). Evolution experiments with yeast show at the genomic level that sexual populations suffer from less clonal interference and fewer deleterious hitchhiker alleles than asexual populations (McDonald et al., 2016). This result corroborates two of the most prevalent hypotheses concerning the advantages of sex. But that, by itself, does not fully unravel the Gordian knot that is the evolution of sex, because it cannot explain the origin of anisogamous sex or even necessarily its maintenance.

With respect to the evolutionary origins of sex, the spread of mutator alleles (Chao & Cox, 1983; Consuegra et al., 2021; Wielgoss et al., 2013) and the spread of mobile conjugative plasmids when present (Bergstrom et al., 2000) in some experiments with bacteria suggest that second-order selection can favor the evolution of genetic mechanisms that promote variation and permit recombination, at least to some extent. Alternatively, recombination-promoting mechanisms could arise and spread because they foster the spread of conjugative plasmids in, of, and for themselves (Rose, 1983).

Most theories for the evolutionary maintenance of anisogamous sex have great difficulty overcoming the many short-term penalties that arise from sex for females (Maynard Smith, 1978; Michod & Levin, 1988). An alternative theory is that females have great difficulty escaping males even when they could undergo successful parthenogenesis in the absence of males (Krieber & Rose, 1986). This theory has never been tested using experimental evolution, but perhaps it could be using the facultatively parthenogenetic *D. mercatorum.*

8.2.4. *Speciation*

Patterns of phenotypic and genetic convergence and divergence in elaborate systems of populations can be produced by experimental evolution, such as those of the Ives-derived populations in the Rose lab. Such complexes of populations provide excellent material for testing alternative theories of speciation, such as those rooted in divergent adaptation to different environmental conditions (e.g., Darwin, 1859) and others that reflect accidental genomic processes (e.g., Rose & Doolittle, 1983). Some initial work along these lines has been done by Cabral (2015) and Robinson et al. (2023).

The origin of citrate utilization in one of the LTEE lines, while in an asexual bacterial system, shows several hallmarks of incipient ecological speciation: the origin of a novel function, coexistence of lineages with and without that function, and subsequent adaptation and divergence of those lineages (Blount et al., 2008, 2020; Turner et al., 2023). In experiments with a virus that can undergo recombination only within host cells, Meyer et al. have gone even further. After discovering that phage λ could infect bacteria through a previously unused receptor (Meyer et al., 2012), they further showed that the evolution of this novelty led to ecological speciation via the reduced opportunity to undergo recombination (Chaikeeratisak et al., 2021; Doud et al., 2024; Meyer et al., 2016).

8.3. Prospects for the Future of Experimental Evolution

8.3.1. *New experimental systems*

Experimental evolution in asexual systems, including the LTEE (Lenski, 2023; Lenski et al., 1991) and other studies with bacteria and yeast (e.g., Bozdag et al., 2023; Gallie et al., 2019; Johnson et al., 2021), can achieve great scientific power with relatively modest levels of space, cost, and personnel. However, the same cannot be said for evolution experiments with outbreeding multicellular organisms. For example, the mouse wheel-running experiments of the Garland lab have suffered from problems of inbreeding depression (see Careau et al., 2013). The Rose laboratory has developed the Ives-derived *Drosophila* populations using thousands of people over the course of more than 40 years, a challenging task for a little-funded field like evolutionary biology.

Promising new systems are being actively developed. The Teotónio lab and others have been developing nematode worms of the genus *Caenorhabditis* as a system for experimental evolution. Although they are multicellular animals, they can be frozen and later revived (like many microbes), and their very small body size allows large populations in small culture vessels (Teotónio et al., 2017). Meanwhile, Burke and others (e.g., Burke, 2023; Burke et al., 2014; Linder et al., 2022) have been using out-crossing yeast in evolution experiments. Both of these systems have great potential for the study of sexual species using experimental evolution.

8.3.2. *New experimental and analytical methods*

While much has been achieved already using experimental evolution with microorganisms, new technological developments indicate even further potential. The advent of robotic transfer systems has enabled the expansion of selection regimes as well as replicates within regimes. For example, Johnson et al. (2021) were able to propagate 205 *Saccharomyces cerevisiae* populations for 10,000 generations over the course of a few years. The LTEE has sequenced hundreds of genomes of both isolated clones (Barrick et al., 2009; Tenaillon et al., 2016) and whole-population samples (Barrick & Lenski, 2009; Good et al., 2017). New approaches using microfluidics allow for the separation and sequencing of individual cells within diverse populations (Zhou et al., 2021), avoiding the need to isolate clones as colonies. Raman-activated gravity-driven single-cell encapsulation (RAGE-seq) allows for the phenotypic sorting of bacteria and their subsequent sequencing (Xu et al., 2020). Such methods should enable both the separation of individuals within experimental populations and the correlation of their individual genomes with the sorted phenotype. Such methods should also allow a finer analysis of genetic draft, as the frequency of linked variants can be determined at specific moments in the population's evolutionary history. Nanopore and other long-read sequencing methods will facilitate genomic analysis of individuals within diverse populations. This technology can also be used to detect copy-number variation in the genomes of both prokaryotic and eukaryotic cells. Evolution experiments have shown that such variation can play an important role in the adaptation of bacteria and yeast to various resources (Blount et al., 2012, 2020; Spealman et al., 2023).

CRISPR-based systems for genomic editing offer another powerful tool for experimental evolution. For example, You et al. (2019) selected yeast for increased cellular heterogeneity. After evolving lineages with increased noise in protein expression, they sequenced genomes and did genetic crosses to identify candidate mutations that might be responsible for that noisy phenotype. They then used the CRISPR-Cas9 method to introduce the candidates into the ancestor and to remove the mutation from the evolved yeast. By doing so, they identified a mutation that was largely responsible for the increased noise in protein expression (You et al., 2019). Other genome-editing methods have been used to move single mutations in the LTEE and then examine their effects on fitness and other phenotypes (e.g., Cooper et al., 2003; Crozat et al., 2005; Khan et al., 2011). We anticipate that CRISPR-based genome editing will prove valuable for analyzing many evolution experiments in the years ahead.

While the genetic analysis of experimental evolution in asexual systems may often be relatively straightforward (e.g., Barrick et al., 2009; Blount et al., 2012; Tenaillon et al., 2012), that is not the case for sexual systems (e.g., Graves et al., 2017; Hillis et al., 2020; Remolina et al., 2012). Nor do the analyses become any simpler or more straightforward by moving to the level of transcriptomes (Barter et al., 2019; Remolina et al., 2012), metabolomes (Phillips, Arnold et al., 2022), or physiology (e.g., Kezos et al., 2023), even though such analyses can be fruitful in elucidating the mechanisms of adaptation. In any case, connecting the complexity at one level of biological organization to another is a Herculean task, perhaps especially in sexual systems.

Accordingly, Mueller et al. (2018), Kezos et al. (2023), and Greenspan et al. (2023) have begun using machine learning to analyze how genetic changes at many sites in the genomes of sexual species undergoing experimental evolution work their way through to influence transcripts and organismal functions. Many kinds of machine-learning tools are available for this purpose. Sorting out and developing the possibilities for this type of analysis will be important work for the next decade or so in the field of experimental evolution.

8.3.3. *New questions*

Finally, we are left with scanning the horizon for future research that may benefit from using experimental evolution. What are new questions that the field has yet to address but on which it could make substantial progress? Lenski (2023) has discussed how the simplicity of the LTEE design is both a scientific strength and a limitation. He placed this tension in the context of an earlier discussion about the strengths and limitations of mathematical models in ecology. Levins (1966) argued that mathematical models can have three desirable features: generality, realism, and precision. Thus far, the LTEE and, more generally, the field of experimental evolution have emphasized precision, which is the result of using simple, well-controlled environments and careful experimental designs.

Despite this simplicity, experimental evolution has delivered general results that correspond to the way evolution often works in the real world. For example, one of the most striking outcomes of evolution experiments in both asexual and sexual systems is parallelism and even convergence (e.g., Deckert-Cruz et al., 2004; Graves et al., 2020, 2017; Lenski & Travisano, 1994; Rainey & Travisano, 1998; Tenaillon et al., 2016).

Parallel and convergent evolution are also often seen in nature (e.g., Hall, 2012; Hume & Martill, 2019; Koga et al., 2010; Losos, 2009; see also Conway Morris, 2003, for a compendium of examples). However, while there are many examples from nature—the numerator—one cannot readily quantify the cases where parallelism and convergence did *not* occur—the denominator (Lenski, 2017a). Thus, there remains considerable interest and even debate concerning the generality of parallel and convergent evolution (Blount et al., 2018; Conway Morris, 2003; Gould, 1989; Losos, 2018). In any case, it is likely that parallelism is especially common in the field of experimental evolution owing to the simplicity of the selective regimes employed, regimes that do not reflect the complexity of the natural world.

Thus, realism with respect to that complexity is generally lacking in experimental evolution. Therefore, one challenge—and an opportunity—going forward is to impose more complex environments, such as ones with additional species (competitors, mutualists, predators, and parasites), more diverse resources, complex spatial structure, and greater variation in abiotic parameters (such as humidity, light, temperature, and toxins). However, there is a bit of evidence that some evolution experiments can predict how organisms will evolve in the real world. One example is an experiment that showed parallel adaptation among fly populations maintained in large enclosures that were exposed to seasonal variation in their environment (Rudman et al., 2022). The experimental design contained elements of both the laboratory (enclosed populations derived from 80 isogenic lines, no migration, provided with a defined fly food) and nature (large populations with ~10^5 flies per cage, variable temperature and humidity, exposure to the natural insect community associated with a small tree in each enclosure, and a microbiome derived from the enclosed conditions). Despite the complexity of the environment, these conditions still resulted in parallel changes in life-history traits (fecundity, egg size, developmental rate), stress resistance (starvation, desiccation, cold tolerance), and genomic variation. More experiments of this type would be valuable to better understand the generality, or limitations, of inferences from experimental evolution.

In nature, bacteria often grow as biofilms on surfaces, and evolution experiments with several species are now being used to better understand biofilms (e.g., Kovács & Dragoš, 2019; Poltak & Cooper, 2011; Rainey & Travisano, 1998). The LTEE was performed with bacteria growing in a well-mixed medium and at a relatively low population density. However, biofilms consist of a densely packed and often multispecies aggregation embedded in a matrix produced by the bacteria themselves. Biofilms occur in multiple forms: submerged (solid-liquid interface), colony (solid-air interface), or pellicle (liquid-air interface). Nutrient and oxygen availability can vary widely within and near biofilms, creating opportunities for the bacteria to adapt to and specialize in diverse niches on a tiny spatial scale (Rainey & Travisano, 1998). Moreover, many biofilms are transient structures, and so bacteria must evolve ways of dispersing between what amount to islands of opportunity (Poltak & Cooper, 2011). Kovács and Dragoš (2019) showed that the evolution of biofilms was influenced by several factors, including coevolution with phage and a tension between cooperator and cheater genotypes with respect to the production of the extracellular matrix. Here, too, we see the expansion of experimental evolution from simple model systems to embracing the complexities of natural microbial communities.

8.4. Envoi and Perspective

Experimental evolution has come a long way from its humble beginnings in the 20th century. It has been deployed to address foundational questions about the roles of mutation, recombination, selection, and drift in driving evolutionary change. Even so, we believe that evolution experiments need to be applied more often to allow strong-inference tests of evolutionary theory going forward. Big questions remain concerning the rates and limits of evolution, the mechanisms of speciation, the origins of multicellularity, and the processes that drive adaptive radiations. We also expect that new applications of experimental evolution will prove useful in arenas such as synthetic biology, for example, by allowing engineered forms to evolve from "rough drafts" to finely tuned organisms.

We hope this volume provides useful guidance to future researchers so they may continue to unravel exactly how, again in Darwin's words (1859), "endless forms most beautiful and most wonderful have been, and are being, evolved."

References

Archer, M. A., Bradley, T. J., Mueller, L. D., & Rose, M. R. (2007). Using experimental evolution to study the physiological mechanisms of desiccation resistance in *Drosophila melanogaster*. *Physiological and Biochemical Zoology, 80*(4), 386–398. https://doi.org/10.1086/518354

Archer, M. A., Phelan, J. P., Beckman, K. A., & Rose, M. R. (2003). Breakdown in correlations during laboratory evolution. II. Selection on stress resistance in *Drosophila* populations. *Evolution, 57*(3), 536–543. https://doi.org/10.1111/j.0014-3820.2003.tb01545.x

Arnold, K. R. (2025). *Genomic Trajectories of Adaptation & the Omics of Aging*. PhD diss., University of California, Irvine. ProQuest ID: Arnold_uci_0030D_19505. Merritt ID: ark:/13030/m5m73bb4. https://escholarship.org/uc/item/04j0v5gx.

Arnold, K. R., & Rose, M. R. (2023). *Conceptual breakthroughs in the evolutionary biology of aging*. Academic Press. New York, NY, USA. https://doi.org/10.1016/C2019-0-03244-X

Atolia, E., Cesar, S., Arjes, H. A., Rajendram, M., Shi, H., Knapp, B. D., Khare, S., Aranda-Díaz, A., Lenski, R. E., & Huang, K. C. (2020). Environmental and physiological factors affecting high-throughput measurements of bacterial growth. *mBio, 11*(5), e01378-20. https://doi.org/10.1128/mbio.01378-20

Avise, J. C. (2014). *Conceptual breakthroughs in evolutionary genetics: A brief history of shifting paradigms*. Academic Press. New York, NY, USA.

Barghi, N., Tobler, R., Nolte, V., Jakšić, A. M., Mallard, F., Otte, K. A., Dolezal, M., Taus, T., Kofler, R., & Schlötterer, C. (2019). Genetic redundancy fuels polygenic adaptation in *Drosophila*. *PLOS Biology, 17*(2), e3000128. https://doi.org/10.1371/journal.pbio.3000128

Barrick, J. E., Blount, Z. D., Lake, D. M., Dwenger, J. H., Chavarria-Palma, J. E., Izutsu, M., & Wiser, M. J. (2023). Daily transfers, archiving populations, and measuring fitness in the long-term evolution experiment with *Escherichia coli*. *Journal of Visualized Experiments, 198*, e65342. https://doi.org/10.3791/65342

Barrick, J. E., & Lenski, R. E. (2009). Genome-wide mutational diversity in an evolving population of *Escherichia coli*. *Cold Spring Harbor Symposia on Quantitative Biology, 74*, 119–129. https://doi.org/10.1101/sqb.2009.74.018

Barrick, J. E., & Lenski, R. E. (2013). Genome dynamics during experimental evolution. *Nature Reviews Genetics, 14*(12), 827–839. https://doi.org/10.1038/nrg3564

Barrick, J. E., Yu, D. S., Yoon, S. H., Jeong, H., Oh, T. K., Schneider, D., Lenski, R. E., & Kim, J. F. (2009). Genome evolution and adaptation in a long-term experiment with *Escherichia coli*. *Nature, 461*(7268), 1243–1247. https://doi.org/10.1038/nature08480

Barter, T. T., Greenspan, Z. S., Phillips, M. A., Mueller, L. D., Rose, M. R., & Ranz, J. M. (2019). *Drosophila* transcriptomics with and without ageing. *Biogerontology, 20*(5), 699–710. https://doi.org/10.1007/s10522-019-09823-4

Basu, A., Tekade, K., Singh, A., Das, P. N., & Prasad, N. G. (2024). Experimental evolution for improved postinfection survival selects for increased disease resistance in *Drosophila melanogaster*. *Evolution, 78*(11), 1831–1843. https://doi.org/10.1093/evolut/qpae116

Baym, M., Lieberman, T. D., Kelsic, E. D., Chait, R., Gross, R., Yelin, I., & Kishony, R. (2016). Spatiotemporal microbial evolution on antibiotic landscapes. *Science, 353*(6304), 1147–1151. https://doi.org/10.1126/science.aag0822

Behringer, M. G., Choi, B. I., Miller, S. F., Doak, T. G., Karty, J. A., Guo, W., & Lynch, M. (2018). *Escherichia coli* cultures maintain stable subpopulation structure during long-term evolution. *Proceedings of the National Academy of Sciences of the United States of America, 115*(20), E4642–E4650. https://doi.org/10.1073/pnas.1708371115

Bell, G. (1984). Evolutionary and nonevolutionary theories of senescence. *The American Naturalist, 124*, 600–603. https://www.jstor.org/stable/2461601

Bell, G. (1988). *Sex and death in protozoa: The history of an obsession.* Cambridge University Press. Cambridge, UK.

Bennett, A. F., & Lenski, R. E. (1993). Evolutionary adaptation to temperature. II. Thermal niches of experimental lines of *Escherichia coli*. *Evolution, 47*(1), 1–12. https://doi.org/10.1111/j.1558-5646.1993.tb01194.x

Bennett, A. F., & Lenski, R. E. (1999). Experimental evolution and its role in evolutionary physiology. *American Zoologist, 39*(2), 346–362. https://doi.org/10.1093/icb/39.2.346

Bennett, A. F., Lenski, R. E., & Mittler, J. E. (1992). Evolutionary adaptation to temperature. I. Fitness responses of *Escherichia coli* to changes in its thermal environment. *Evolution, 46*(1), 16–30. https://doi.org/10.1111/j.1558-5646.1992.tb01981.x

Bergstrom, C. T., Lipsitch, M., & Levin, B. R. (2000). Natural selection, infectious transfer and the existence conditions for bacterial plasmids. *Genetics, 155*(4), 1505–1519. https://doi.org/10.1093/genetics/155.4.1505

Blount, Z. D., Barrick, J. E., Davidson, C. J., & Lenski, R. E. (2012). Genomic analysis of a key innovation in an experimental *Escherichia coli* population. *Nature, 489*(7417), 513–518. https://doi.org/10.1038/nature11514

Blount, Z. D., Borland, C. Z., & Lenski, R. E. (2008). Historical contingency and the evolution of a key innovation in an experimental population of *Escherichia coli. Proceedings of the National Academy of Sciences of the United States of America, 105*(23), 7899–7906. https://doi.org/10.1073/pnas.0803151105

Blount, Z. D., Lenski, R. E., & Losos, J. B. (2018). Contingency and determinism in evolution: Replaying life's tape. *Science, 362*(6415), eaam5979. https://doi.org/10.1126/science.aam5979

Blount, Z. D., Maddamsetti, R., Grant, N. A., Ahmed, S. T., Jagdish, T., Baxter, J. A., Sommerfeld, B. A., Tillman, A., Moore, J., Slonczewski, J. L., Barrick, J. E., & Lenski, R. E. (2020). Genomic and phenotypic evolution of *Escherichia coli* in a novel citrate-only resource environment. *eLife, 9*, e55414. https://doi.org/10.7554/eLife.55414

Blundell, J. R., Schwartz, K., Francois, D., Fisher, D. S., Sherlock, G., & Levy, S. F. (2019). The dynamics of adaptive genetic diversity during the early stages of clonal evolution. *Nature Ecology & Evolution, 3*(2), 293–301. https://doi.org/10.1038/s41559-018-0758-1

Bohannan, B. J. M., & Lenski, R. E. (2000). Linking genetic change to community evolution: insights from studies of bacteria and bacteriophage. *Ecology Letters, 3*(4), 362–377. https://doi.org/10.1046/j.1461-0248.2000.00161.x

Bouma, J. E., & Lenski, R. E. (1988). Evolution of a bacteria/plasmid association. *Nature, 335*(6188), 351–352. https://doi.org/10.1038/335351a0

Bowler, P. J. (1989). *The Mendelian revolution: The emergence of hereditarian concepts in modern science and society.* Johns Hopkins University Press. Baltimore, MD.

Boyd, S. M., Rhinehardt, K., Ewunkem, J. A., Harrison, S. H., Thomas, M. D., & Graves, J. L. Jr. (2022). Experimental evolution of copper resistance in *Escherichia coli* produces evolutionary trade-offs in the antibiotics chloramphenicol, bacitracin, and sulfonamide. *Antibiotics*, *11*(6), 711. https://doi.org/10.3390/antibiotics11060711

Bozdag, G. O., Zamani Dahaj, S. A., Day, T. C., Kahn, P. C., Burnetti, A. J., Lac, D. T., Tong, K., Conlin, P. L., Balwani, A. H., Dyer, E. L., Yunker, P. J., & Ratcliff, W. C. (2023). De novo evolution of macroscopic multicellularity. *Nature*, *617*(7962), 747–754. https://doi.org/10.1038/s41586-023-06052-1

Brawand, D., Wagner, C. E., Li, Y. I., Malinsky, M., Keller, I., Fan, S., Simakov, O., Ng, A. Y., Lim, Z. W., Bezault, E., Turner-Maier, J., Johnson, J., Alcazar, R., Noh, H. J., Russell, P., Aken, B., Alföldi, J., Amemiya, C., Azzouzi, N., . . . Di Palma, F. (2014). The genomic substrate for adaptive radiation in African cichlid fish. *Nature*, *513*(7518), 375–381. https://doi.org/10.1038/nature13726

Burke, M. K. (2012). How does adaptation sweep through the genome? Insights from long-term selection experiments. *Proceedings of the Royal Society B: Biological Sciences*, *279*(1749), 5029–5038. https://doi.org/10.1098/rspb.2012.0799

Burke, M. K. (2023). Embracing complexity: Yeast evolution experiments featuring standing genetic variation. *Journal of Molecular Evolution*, *91*(3), 281–292. https://doi.org/10.1007/s00239-023-10094-4

Burke, M. K., Barter, T. T., Cabral, L. G., Kezos, J. N., Phillips, M. A., Rutledge, G. A., Phung, K. H., Chen, R. H., Nguyen, H. D., Mueller, L. D., & Rose, M. R. (2016). Rapid divergence and convergence of life-history in experimentally evolved *Drosophila melanogaster*. *Evolution*, *70*(9), 2085–2098. https://doi.org/10.1111/evo.13006

Burke, M. K., Dunham, J. P., Shahrestani, P., Thornton, K. R., Rose, M. R., & Long, A. D. (2010). Genome-wide analysis of a long-term evolution experiment with *Drosophila*. *Nature*, *467*(7315), 587–590. https://doi.org/10.1038/nature09352

Burke, M. K., Liti, G., & Long, A. D. (2014). Standing genetic variation drives repeatable experimental evolution in outcrossing populations of *Saccharomyces cerevisiae*. *Molecular Biology and Evolution*, *31*(12), 3228–3239. https://doi.org/10.1093/molbev/msu256

Burke, M. K., & Rose, M. R. (2009). Experimental evolution with *Drosophila*. *American Journal of Physiology: Regulatory, Integrative and Comparative Physiology*, *296*(6), R1847–R1854. https://doi.org/10.1152/ajpregu.90551.2008

Burmeister, A. R., Lenski, R. E., & Meyer, J. R. (2016). Host coevolution alters the adaptive landscape of a virus. *Proceedings of the Royal Society B: Biological Sciences, 283*, 20161528. https://doi.org/10.1098/rspb.2016.1528

Burny, C., Nolte, V., Dolezal, M., & Schlötterer, C. (2021). Highly parallel genomic selection response in replicated *Drosophila melanogaster* populations with reduced genetic variation. *Genome Biology and Evolution, 13*(11), evab239. https://doi.org/10.1093/gbe/evab239

Cabral, L. G. (2015). *Experimental tests of speciation mechanisms in Drosophila melanogaster* (Doctoral dissertation). University of California, Irvine, CA, USA.

Card, K. J., LaBar, T., Gomez, J. B., & Lenski, R. E. (2019). Historical contingency in the evolution of antibiotic resistance after decades of relaxed selection. *PLOS Biology, 17*(10), e3000397. https://doi.org/10.1371/journal.pbio.3000397

Card, K. J., Thomas, M. D., Graves, J. L. Jr., Barrick, J. E., & Lenski, R. E. (2021). Genomic evolution of antibiotic resistance is contingent on genetic background following a long-term experiment with *Escherichia coli*. *Proceedings of the National Academy of Sciences of the United States of America, 118*(5), e2016886118. https://doi.org/10.1073/pnas.2016886118

Careau, V., Wolak, M. E., Carter, P. A., & Garland, T. (2013). Limits to behavioral evolution: The quantitative genetics of a complex trait under directional selection. *Evolution, 67*(11), 3102–3119. https://doi.org/10.1111/evo.12200

Carey, J. R., Liedo, P., Orozco, D., & Vaupel, J. W. (1992). Slowing of mortality rates at older ages in large medfly cohorts. *Science, 258*(5081), 457–461. https://doi.org/10.1126/science.1411540

Carolus, H., Pierson, S., Muñoz, J. F., Subotić, A., Cruz, R. B., Cuomo, C. A., & Van Dijck, P. (2021). Genome-wide analysis of experimentally evolved *Candida auris* reveals multiple novel mechanisms of multidrug resistance. *mBio, 12*(2), e03333-20. https://doi.org/10.1128/mBio.03333-20

Castle, W. E., & Phillips, J. C. (1914). *Piebald rats and selection; an experimental test of the effectiveness of selection and of the theory of gametic purity in Mendelian crosses.* Carnegie Institution of Washington. Washington, DC, USA

Chaikeeratisak, V., Birkholz, E. A., Prichard, A. M., Egan, M. E., Mylvara, A., Nonejuie, P., Nguyen, K. T., Sugie, J., Meyer, J. R., & Pogliano, J. (2021). Viral speciation through subcellular genetic isolation and virogenesis incompatibility. *Nature Communications, 12*(1), Article 342. https://doi.org/10.1038/s41467-020-20575-5

Chao, L., & Cox, E. C. (1983). Competition between high and low mutating strains of *Escherichia coli*. *Evolution*, *37*(1), 125–134. https://doi.org/10.1111/j.1558-5646.1983.tb05521.x

Chao, L., Levin, B. R., & Stewart, F. M. (1977). A complex community in a simple habitat: an experimental study with bacteria and phage. *Ecology*, *58*(2), 369–378. https://doi.org/10.2307/1935611

Charlesworth, B. (1980). *Evolution in age-structured populations*. Cambridge University Press. Cambridge, UK.

Charlesworth, B. (1994). *Evolution in age-structured populations* (2nd ed.). Cambridge University Press. Cambridge, UK. https://doi.org/10.1017/CBO9780511525711

Charlesworth, B. (2001). Patterns of age-specific means and genetic variances of mortality rates predicted by the mutation-accumulation theory of ageing. *Journal of Theoretical Biology*, *210*(1), 47–65. https://doi.org/10.1006/jtbi.2001.2296

Charlesworth, B., & Hughes, K. A. (1996). Age-specific inbreeding depression and components of genetic variance in relation to the evolution of senescence. *Proceedings of the National Academy of Sciences of the United States of America*, *93*(12), 6140–6145. https://doi.org/10.1073/pnas.93.12.6140

Charlesworth, B., & Williamson, J. A. (1975). The probability of survival of a mutant gene in an age-structured population and implications for the evolution of life-histories. *Genetical Research*, *26*(1), 1–10. https://doi.org/10.1017/S0016672300015792

Chaudhuri, R. R., & Henderson, I. R. (2012). The evolution of the *Escherichia coli* phylogeny. *Infection, Genetics and Evolution*, *12*(2), 214–226. https://doi.org/10.1016/j.meegid.2012.01.005

Chippindale, A. K., Alipaz, J. A., Chen, H. W., & Rose, M. R. (1997). Experimental evolution of accelerated development in *Drosophila*. I. Developmental speed and larval survival. *Evolution*, *51*(5), 1536–1551. https://doi.org/10.1111/j.1558-5646.1997.tb01477.x

Chippindale, A. K., Leroi, A. M., Kim, S. B., & Rose, M. R. (1993). Phenotypic plasticity and selection in *Drosophila* life-history evolution. I. Nutrition and the cost of reproduction. *Journal of Evolutionary Biology*, *6*(2), 171–193. https://doi.org/10.1046/j.1420-9101.1993.6020171.x

Clare, M. J., & Luckinbill, L. S. (1985). The effects of gene–environment interaction on the expression of longevity. *Heredity*, *55*, 19–26. https://doi.org/10.1038/hdy.1985.67

Cohan, F. M. (1994). Genetic exchange and evolutionary divergence in prokaryotes. *Trends in Ecology & Evolution*, *9*(5), 175–180. https://doi.org/10.1016/0169-5347(94)90081-7

Comfort, A. (1953). Absence of a Lansing effect in *Drosophila subobscura*. *Nature*, *172*(4367), 83–84. https://doi.org/10.1038/172083a0

Comfort, A. (1979). *The biology of senescence* (3rd ed.). Elsevier. London, UK.

Conery, M., & Grant, S. F. A. (2023). Human height: A model common complex trait. *Annals of Human Biology*, *50*(1), 258–266. https://doi.org/10.1080/03014460.2023.2215546

Consuegra, J., Gaffé, J., Lenski, R. E., Hindré, T., Barrick, J. E., Tenaillon, O., & Schneider, D. (2021). IS-mediated mutations both promote and constrain evolvability during a long-term experiment with bacteria. *Nature Communications 12*(1), 980. https://doi.org/10.1038/s41467-021-21210-7

Conway Morris, S. (2003). *Life's solution: Inevitable humans in a lonely universe*. Cambridge University Press. Cambridge, UK.

Cooper, T. F., Remold, S. K., Lenski, R. E., & Schneider, D. (2008). Expression profiles reveal parallel evolution of epistatic interactions involving the CRP regulon in *Escherichia coli*. *PLoS Genetics*, *4*(2), e35. https://doi.org/10.1371/journal.pgen.0040035

Cooper, T. F., Rozen, D. E., & Lenski, R. E. (2003). Parallel changes in gene expression after 20,000 generations of evolution in *Escherichia coli*. *Proceedings of the National Academy of Sciences of the United States of America*, *100*(3), 1072–1077. https://doi.org/10.1073/pnas.0334340100

Cooper, V. S., Bennett, A. F., & Lenski, R. E. (2001). Evolution of thermal dependence of growth rate of *Escherichia coli* populations during 20,000 generations in a constant environment. *Evolution*, *55*(5), 889–896. https://doi.org/10.1111/j.0014-3820.2001.tb00606.x

Cooper, V. S., & Lenski, R. E. (2000). The population genetics of ecological specialization in evolving *Escherichia coli* populations. *Nature*, *407*(6805), 736–739. https://doi.org/10.1038/35037572

Couce, A., Limdi, A., Magnan, M., Owen, S. V., Herren, C. M., Lenski, R. E., Tenaillon, O., & Baym, M. (2024). Changing fitness effects of mutations through long-term bacterial evolution. *Science*, *383*(6681), eadd1417. https://doi.org/10.1126/science.add1417

Crozat, E., Philippe, N., Lenski, R. E., Geiselmann, J., & Schneider, D. (2005). Long-term experimental evolution in *Escherichia coli*. XII. DNA topology as a key target of selection. *Genetics*, *169*(2), 523–532. https://doi.org/10.1534/genetics.104.035717

Csilléry, K., Rodríguez-Verdugo, A., Rellstab, C., & Guillaume, F. (2018). Detecting the genomic signal of polygenic adaptation and the role of epistasis in evolution. *Molecular Ecology*, *27*(3), 606–612. https://doi.org/10.1111/mec.14499

Curtsinger, J. W., Fukui, H. H., Townsend, D. R., & Vaupel, J. W. (1992). Demography of genotypes: Failure of the limited life-span paradigm in *Drosophila melanogaster*. *Science*, *258*(5081), 461–463. https://doi.org/10.1126/science.1411541

Darwin, C. (1859). *On the origin of species by means of natural selection, or the preservation of favoured races in the struggle for life*. John Murray. London, UK.

de Visser, J. A. G. M., Zeyl, C. W., Gerrish, P. J., Blanchard, J. L., & Lenski, R. E. (1999). Diminishing returns from mutation supply rate in asexual populations. *Science*, *283*(5400), 404–406. https://doi.org/10.1126/science.283.5400.404

Deatherage, D. E., & Barrick, J. E. (2014). Identification of mutations in laboratory-evolved microbes from next-generation sequencing data using breseq. In L. Sun & W. Shou(eds), *Engineering and analyzing multicellular systems: Methods in molecular biology* (Vol. 1151, pp. 165–188). Humana Press. New York, N.Y. USA. https://doi.org/10.1007/978-1-4939-0554-6_12

Deatherage, D. E., & Barrick, J. E. (2021). High throughput characterization of mutations in genes that drive clonal evolution using multiplex adaptome capture sequencing. *Cell Systems*, *12*(12), 1187–1200.e4. https://doi.org/10.1016/j.cels.2021.08.011

Deatherage, D. E., Kepner, J. L., Bennett, A. F., Lenski, R. E., & Barrick, J. E. (2017). Specificity of genome evolution in experimental populations of *Escherichia coli* evolved at different temperatures. *Proceedings of the National Academy of Sciences of the United States of America*, *114*(10), E1904–E1912. https://doi.org/10.1073/pnas.1616132114

Deckert-Cruz, D. J., Matzkin, L. M., Graves, J. L., & Rose, M. R. (2004). Electrophoretic analysis of Methuselah flies from multiple species. In M. R. Rose, H. B. Passananti & M. Matos (Eds.), *Methuselah flies* (pp. 237–248). World Scientific Publishing. Singapore.

Deckert-Cruz, D. J., Tyler, R. H., Landmesser, J. E., & Rose, M. R. (1997). Allozymic differentiation in response to laboratory demographic selection of *Drosophila melanogaster*. *Evolution*, *51*(3), 865–872. https://doi.org/10.1111/j.1558-5646.1997.tb03668.x

DeLong, J. P., Okie, J. G., Moses, M. E., Sibly, R. M., & Brown, J. H. (2010). Shifts in metabolic scaling, production, and efficiency across major evolutionary transitions of life. *Proceedings of the National Academy of Sciences*

of the United States of America, 107(29), 12941–12945. https://doi.
org/10.1073/pnas.1007783107

Djawdan, M., Chippindale, A. K., Rose, M. R., & Bradley, T. J. (1998). Metabolic
reserves and evolved stress resistance in *Drosophila melanogaster.
Physiological and Biochemical Zoology, 71*(5), 584–594. https://doi.
org/10.1086/515963

Dobzhansky, T. (1973). Nothing in biology makes sense except in the light of
evolution. *The American Biology Teacher, 35*(3), 125–129. https://doi.
org/10.2307/4444260

Dodd, D. M. B. (1989). Reproductive isolation as a consequence of adaptive
divergence in *Drosophila pseudoobscura. Evolution, 43*(6), 1308–1311.
https://doi.org/10.1111/j.1558-5646.1989.tb02577.x

Doud, M. B., Gupta, A., Li, V., Medina, S. J., De La Fuente, C. A., & Meyer, J.
R. (2024). Competition-driven eco-evolutionary feedback reshapes bacteriophage lambda's fitness landscape and enables speciation. *Nature
Communications, 15*(1), Article 863. https://doi.org/10.1038/s41467-024-
45008-5

Dykhuizen, D. E., & Dean, A. M. (2009). Experimental evolution from the bot-
tom up. In T. D. Garland & M. R. Rose (Eds.), *Experimental evolution:
Concepts, methods, and applications of selection experiments* (pp. 67–87).
University of California Press. Berkeley, CA, USA

Elena, S. F., Cooper, V. S., & Lenski, R. E. (1996). Punctuated evolution caused
by selection of rare beneficial mutations. *Science, 272*(5269), 1802–1804.
https://doi.org/10.1126/science.272.5269.1802

Elena, S. F., & Lenski, R. E. (1997). Long-term experimental evolution in
Escherichia coli. VII. Mechanisms maintaining genetic variability within
populations. *Evolution, 51*(4), 1058–1067. https://doi.org/10.1111/j.1558-
5646.1997.tb03953.x

Ellington, A. D., & Szostak, J. W. (1990). In vitro selection of RNA molecules
that bind specific ligands. *Nature, 346*(6287), 818–822. https://doi.
org/10.1038/346818a0

Ewunkem, J. A., Rodgers, L., Campbell, D., Staley, C., Subedi, K., Boyd, S., &
Graves, J. L. Jr. (2021). Experimental evolution of magnetite nanoparticle
resistance in *Escherichia coli. Nanomaterials (Basel), 11*(3), 790. https://
doi.org/10.3390/nano11030790

Falconer, D. S. (1960). *Introduction to quantitative genetics.* Oliver & Boyd.
Edinburgh, UK.

Falconer, D. S., & Mackay, T. F. C. (1996). *Introduction to quantitative genetics*
(4th ed.). Longman (Pearson Education). Essex, England.

Favate, J. S., Liang, S., Cope, A. L., Yadavalli, S. S., & Shah, P. (2022). The landscape of transcriptional and translational changes over 22 years of bacterial adaptation. *eLife*, *11*, e81979. https://doi.org/10.7554/eLife.81979

Feder, M. E., Bennett, A. F., & Huey, R. B. (2000). Evolutionary physiology. *Annual Review of Ecology and Systematics*, *31*, 315–341. https://doi.org/10.1146/annurev.ecolsys.31.1.315

Fiegna, F., & Velicer, G. J. (2003). Competitive fates of bacterial social parasites: Persistence and self-induced extinction of *Myxococcus xanthus* cheaters. *Proceedings of the Royal Society B: Biological Sciences*, *270*(1523), 1527–1534. https://doi.org/10.1098/rspb.2003.2387

Fiegna, F., Yu, Y. T. N., Kadam, S. V., & Velicer, G. J. (2006). Evolution of an obligate social cheater to a superior cooperator. *Nature*, *441*(7090), 310–314. https://doi.org/10.1038/nature04677

Firman, R. C., & Simmons, L. W. (2011). Experimental evolution of sperm competitiveness in a mammal. *BMC Evolutionary Biology*, *11*, Article 19. https://doi.org/10.1186/1471-2148-11-19

Fisher, R. A. (1930). *The genetical theory of natural selection*. Clarendon Press. Oxford, UK.

Fleming, J. E., Spicer, G. S., Garrison, R. C., & Rose, M. R. (1993). Two-dimensional protein electrophoretic analysis of postponed aging in *Drosophila*. *Genetica*, *91*(1–3), 183–198. https://doi.org/10.1007/BF01435997

Ford, E. B. (1964). *Ecological genetics*. Chapman & Hall. London, UK.

Fragata, I., Lopes-Cunha, M., Bárbaro, M., Kellen, B., Lima, M., Santos, M. A., Faria, G. S., Santos, M., Matos, M., & Simões, P. (2014). How much can history constrain adaptive evolution? A real-time evolutionary approach of inversion polymorphisms in *Drosophila subobscura*. *Journal of Evolutionary Biology*, *27*(12), 2727–2738. https://doi.org/10.1111/jeb.12533

Fragata, I., Simões, P., Lopes-Cunha, M., Lima, M., Kellen, B., Bárbaro, M., Santos, J., Rose, M. R., Santos, M., & Matos, M. (2014). Laboratory selection quickly erases historical differentiation. *PLoS One*, *9*(5), e96227. https://doi.org/10.1371/journal.pone.0096227

Fukami, T., Beaumont, H., Zhang, X. X., & Rainey, P. B. (2007). Immigration history controls diversification in experimental adaptive radiation. *Nature*, *446*(7134), 436–439. https://doi.org/10.1038/nature05629

Gallie, J., Bertels, F., Remigi, P., Ferguson, G. C., Nestmann, S., & Rainey, P. B. (2019). Repeated phenotypic evolution by different genetic routes in *Pseudomonas fluorescens* SBW25. *Molecular Biology and Evolution*, *36*(5), 1071–1085. https://doi.org/10.1093/molbev/msz040

Garland, T. Jr., Morgan, M. T., Swallow, J. G., Rhodes, J. S., Girard, I., Belter, J. G., & Carter, P. A. (2002). Evolution of a small-muscle polymorphism in lines of house mice selected for high activity levels. *Evolution, 56*(6), 1267–1275. https://doi.org/10.1111/j.0014-3820.2002.tb01437.x

Garland, T. Jr., & Rose, M. R. (Eds.). (2009). *Experimental evolution: Concepts, methods, and applications of selection.* University of California Press. Berkeley, USA. https://doi.org/10.1525/california/9780520247666.001.0001

Geddes, P., & Thomson, J. A. (1908). *The evolution of sex.* W. Scott. New York, NY.

Gerrish, P. J., & Lenski, R. E. (1998). The fate of competing beneficial mutations in an asexual population. *Genetica, 102*(1), 127–144. https://doi.org/10.1023/A:1017067816551

Gibbs, A. G. (1999). Laboratory selection for the comparative physiologist. *Journal of Experimental Biology, 202*(20), 2709–2718. https://doi.org/10.1242/jeb.202.20.2709

Gibbs, A. G., Chippindale, A. K., & Rose, M. R. (1997). Physiological mechanisms of evolved desiccation resistance in *Drosophila melanogaster. Journal of Experimental Biology, 200*(12), 1821–1832. https://doi.org/10.1242/jeb.200.12.1821

Goddard, M. R., Godfray, H. C. J., & Burt, A. (2005). Sex increases the efficacy of natural selection in experimental yeast populations. *Nature, 434*(7035), 636–640. https://doi.org/10.1038/nature03405

Good, B. H., McDonald, M. J., Barrick, J. E., Lenski, R. E., & Desai, M. M. (2017). The dynamics of molecular evolution over 60,000 generations. *Nature, 551*(7678), 45–50. https://doi.org/10.1038/nature24287

Gould, S. J. (1989). *Wonderful life: The Burgess Shale and the nature of history.* W. W. Norton & Company. New York, NY, EUA.

Gould, S. J. (1996). *Full house: The spread of excellence from Plato to Darwin.* Three Rivers Press. New York, NY.

Gould, S. J., & Eldredge, N. (1993). Punctuated equilibrium comes of age. *Nature, 366*(6452), 223–227. https://doi.org/10.1038/366223a0

Grant, N. A., Abdel Magid, A., Franklin, J., Dufour, Y., & Lenski, R. E. (2021). Changes in cell size and shape during 50,000 generations of experimental evolution with *Escherichia coli. Journal of Bacteriology, 203*(10), e00469-20. https://doi.org/10.1128/JB.00469-20

Grant, P. R., & Grant, B. R. (2014). *40 years of evolution: Darwin's finches on Daphne Major Island.* Princeton University Press. Princeton, NJ.

Graves, J. L. (2023). Genetics and American science. In T. Kneeland (Ed.), *The Routledge history of American science* (pp. 357–370). Routledge. London, UK.

Graves, J. L. Jr., Ewunkem, A. J., Thomas, M. D., Han, J., Rhinehardt, K. L., Boyd, S., Edmondson, R., Jeffers-Francis, L., & Harrison, S. H. (2020). Experimental evolution of metal resistance in bacteria. In W. Banzhaf, R. Chait, & A. (Eds.), *Evolution in Action — Past, Present, and Future* (pp. 91–106). Springer International Publishing. Cham, Switzerland. https://doi.org/10.1007/978-3-030-39831-6_9

Graves, J. L. Jr., Ewunkem, A. J., Ward, J., Staley, C., Thomas, M. D., Rhinehardt, K. L., Han, J., & Harrison, S. H. (2019). Experimental evolution of gallium resistance in *Escherichia coli. Evolution, Medicine, and Public Health, 2019*(1), 169–180. https://doi.org/10.1093/emph/eoz025

Graves, J. L. Jr., Hertweck, K. L., Phillips, M. A., Han, M. V., Cabral, L. G., Barter, T. T., Greer, L. F., Burke, M. K., Mueller, L. D., & Rose, M. R. (2017). Genomics of parallel experimental evolution in *Drosophila. Molecular Biology and Evolution, 34*(4), 831–842. https://doi.org/10.1093/molbev/msw282

Graves, J. L. Jr., Luckinbill, L. S., & Nichols, A. (1988). Flight duration and wing beat frequency in long- and short-lived *Drosophila melanogaster. Journal of Insect Physiology, 34*(11), 1021–1026. https://doi.org/10.1016/0022-1910(88)90201-6

Graves, J. L. Jr., Tajkarimi, M., Cunningham, Q., Campbell, A., Nonga, H., Harrison, S. H., & Barrick, J. E. (2015). Rapid evolution of silver nanoparticle resistance in *Escherichia coli. Frontiers in Genetics, 6*, 42. https://doi.org/10.3389/fgene.2015.00042

Greenspan, Z. S., Barter, T. T., Phillips, M. A., Ranz, J. M., Rose, M. R., & Mueller, L. D. (2023). Genomewide architecture of adaptation in experimentally evolved *Drosophila* characterized by widespread pleiotropy. *Journal of Genetics, 103*(1), Article8. https://doi.org/10.1007/s12041-023-01460-8

Greenwood, M., & Irwin, J. O. (1939). The biostatistics of senility. *Human Biology, 11*(1), 1–23. https://www.jstor.org/stable/41447403

Grosberg, R. K., & Strathmann, R. R. (2007). The evolution of multicellularity: A minor major transition? *Annual Review of Ecology, Evolution, and Systematics, 38*(1), 621–654. https://doi.org/10.1146/annurev.ecolsys.36.102403.114735

Großkopf, T., Consuegra, J., Gaffé, J., Willison, J. C., Lenski, R. E., Soyer, O. S., & Schneider, D. (2016). Metabolic modelling in a dynamic evolutionary framework predicts adaptive diversification of bacteria in a long-term evolution experiment. *BMC Evolutionary Biology, 16*(1), Article 163. https://doi.org/10.1186/s12862-016-0733-x

Guirao-Rico, S., & González, J. (2021). Benchmarking the performance of Pool-seq SNP callers using simulated and real sequencing data. *Molecular Ecology Resources*, *21*(4), 1216–1229. https://doi.org/10.1111/1755-0998.13343

Gupta, A., Zaman, L., Strobel, H. M., Gallie, J., Burmeister, A. R., Kerr, B., Tamar, E. S., Kishony, R., & Meyer, J. R. (2022). Host-parasite coevolution promotes innovation through deformations in fitness landscapes. *eLife*, *11*, e76162. https://doi.org/10.7554/eLife.76162

Hall, B. K. (2012). Parallelism, deep homology, and evo-devo. *Evolution & Development*, *14*(1), 29–33. https://doi.org/10.1111/j.1525-142X.2011.00520.x

Hamilton, W. D. (1966). The moulding of senescence by natural selection. *Journal of Theoretical Biology*, *12*(1), 12–45. https://doi.org/10.1016/0022-5193(66)90184-6

Hamilton, W. D. (1980). Sex versus non-sex versus parasite. *Oikos*, *35*, 282–290. https://doi.org/10.2307/3544435

Harari, Y., Ram, Y., Rappoport, N., Hadany, L., & Kupiec, M. (2018). Spontaneous changes in ploidy are common in yeast. *Current Biology*, *28*(6), 825–835.e4. https://doi.org/10.1016/j.cub.2018.01.062

Harshman, L. G., & Hoffmann, A. A. (2000). Laboratory selection experiments using *Drosophila*: What do they really tell us? *Trends in Ecology & Evolution*, *15*(1), 32–36. https://doi.org/10.1016/S0169-5347(99)01756-5

Herron, M. D., Conlin, P. L., & Ratcliff, W. C. (Eds.). (2022). *The evolution of multicellularity*. CRC Press. Boca Raton, FL.

Hickey, D. A., & Rose, M. R. (1988). The role of gene transfer in the evolution of eukaryotic sex. In R. E. Michod & B. R. Levin (Eds.), *The evolution of sex* (pp. 161–175). Sinauer Associates. Sunderland, MA.

Higgs, P. G. (2017). Chemical evolution and the evolutionary definition of life. *Journal of Molecular Evolution*, *84*(5–6), 225–235. https://doi.org/10.1007/s00239-017-9799-3

Hillesland, K. L., Velicer, G. J., & Lenski, R. E. (2009). Experimental evolution of a microbial predator's ability to find prey. *Proceedings of the Royal Society B: Biological Sciences*, *276*, 459–467. https://doi.org/10.1098/rspb.2008.1098

Hillis, D. A., & Garland, T. (2023). Multiple solutions at the genomic level in response to selective breeding for high locomotor activity. *Genetics*, *223*(1), iyac165. https://doi.org/10.1093/genetics/iyac165

Hillis, D. A., Yadgary, L., Weinstock, G. M., De Villena, F. P. M., Pomp, D., Fowler, A. S., Xu, S., Chan, F., & Garland, T. (2020). Genetic basis of aerobically supported voluntary exercise: Results from a selection experiment with house mice. *Genetics*, *216*(3), 781–804. https://doi.org/10.1534/genetics.120.303668

Hoffmann, A. A., Hallas, R., Sinclair, C., & Partridge, L. (2001). Rapid loss of stress resistance in *Drosophila melanogaster* under adaptation to laboratory culture. *Evolution*, *55*(2), 436–438. https://doi.org/10.1111/j.0014-3820.2001.tb01305.x

Hume, J. P., & Martill, D. (2019). Repeated evolution of flightlessness in *Dryolimnas* rails (Aves: Rallidae) after extinction and recolonization on Aldabra. *Zoological Journal of the Linnean Society*, *186*(3), 666–672. https://doi.org/10.1093/zoolinnean/zlz018

Huxley, J. S. (1942). *Evolution, the modern synthesis*. Allen & Unwin. London, UK.

Ishiguro, N., Oka, C., Hanazawa, Y., & Sato, G. (1979). Plasmids in *Escherichia coli* controlling citrate-utilizing ability. *Applied and Environmental Microbiology*, *38*(5), 956–964. https://doi.org/10.1128/aem.38.5.956-964.1979

Ivanov, D. K., Escott-Price, V., Ziehm, M., Magwire, M. M., Mackay, T. F., Partridge, L., & Thornton, J. M. (2015). Longevity GWAS using the *Drosophila* Genetic Reference Panel. *The Journals of Gerontology: Series A*, *70*(12), 1470–1478. https://doi.org/10.1093/gerona/glv047

Ives, P. T. (1970). Further genetic studies of the South Amherst population of *Drosophila melanogaster*. *Evolution*, *24*(3), 507–518. https://doi.org/10.1111/j.1558-5646.1970.tb01785.x

Izutsu, M., & Lenski, R. E. (2022). Experimental test of the contributions of initial variation and new mutations to adaptive evolution in a novel environment. *Frontiers in Ecology and Evolution*, *10*, Article 958406. https://doi.org/10.3389/fevo.2022.958406

Izutsu, M., Lake, D. M., Matson, Z. W. D., Dodson, J. P., & Lenski, R. E. (2024). Effects of periodic bottlenecks on the dynamics of adaptive evolution in microbial populations. *Microbiology*, *170*(9), 001494. https://doi.org/10.1099/mic.0.001494

Jacob, F. (1977). Evolution and tinkering. *Science*, *196*(4295), 1161–1166. https://doi.org/10.1126/science.860134

Jacobeen, S., Pentz, J. T., Graba, E. C., Brandys, C. G., Ratcliff, W. C., & Yunker, P. J. (2018). Cellular packing, mechanical stress and the evolution of multicellularity. *Nature Physics*, *14*(3), 286–290. https://doi.org/10.1038/s41567-017-0002-y

Jagdish, T. (2023). *The dynamics of fitness and pleiotropy in a long-term evolution experiment with Escherichia coli* (Doctoral dissertation). Harvard University.

Jeje, O., Ewunkem, J. A., Jeffers-Francis, L. K., & Graves, J. L. (2023). Serving two masters: Effect of *Escherichia coli* dual resistance on antibiotic susceptibility. *Antibiotics, 12*(3), 603. https://doi.org/10.3390/antibiotics12030603

Jeong, H., Barbe, V., Lee, C. H., Vallenet, D., Yu, D. S., Choi, S.-H., Couloux, A., Lee, S.-W., Yoon, S. H., Cattolico, L., Hur, C.-G., Park, H.-S., Ségurens, B., Kim, S. C., Oh, T. K., Lenski, R. E., Studier, F. W., Daegelen, P., & Kim, J. F. (2009). Genome sequences of *Escherichia coli* B strains REL606 and BL21(DE3). *Journal of Molecular Biology, 394*(4), 644–652. https://doi.org/10.1016/j.jmb.2009.09.052

Johnson, M. S., Gopalakrishnan, S., Goyal, J., Dillingham, M. E., Bakerlee, C. W., Humphrey, P. T., Jagdish, T., Jerison, E. R., Kosheleva, K., Lawrence, K. R., Min, J., Moulana, A., Phillips, A. M., Piper, J. C., Purkanti, R., Rego Costa, A., McDonald, M. J., Nguyen Ba, A. N., & Desai, M. M. (2021). Phenotypic and molecular evolution across 10,000 generations in laboratory budding yeast populations. *eLife, 10*, e63910. https://doi.org/10.7554/eLife.63910

Kassen, R., Buckling, A., Bell, G., & Rainey, P. B. (2000). Diversity peaks at intermediate productivity in a laboratory microcosm. *Nature, 406*(6799), 508–512. https://doi.org/10.1038/35020060

Kawecki, T. J., Lenski, R. E., Ebert, D., Hollis, B., Olivieri, I., & Whitlock, M. C. (2012). Experimental evolution. *Trends in Ecology & Evolution, 27*(10), 547–560. https://doi.org/10.1016/j.tree.2012.06.001

Kellermann, V., Hoffmann, A. A., Kristensen, T. N., Moghadam, N. N., & Loeschcke, V. (2015). Experimental evolution under fluctuating thermal conditions does not reproduce patterns of adaptive clinal differentiation in *Drosophila melanogaster*. *The American Naturalist, 186*(5), 582–593. https://doi.org/10.1086/683252

Kettlewell, H. B. (1959). Darwin's missing evidence. *Scientific American, 200*(3), 48–53. https://doi.org/10.1038/scientificamerican0359-48

Kezos, J. N., Barter, T. T., Phillips, M. A., Cabral, L. G., Greenspan, Z. S., Arnold, K. R., Azatian, G., Buenrostro, J., Bhangoo, P. S., Khong, A., Reyes, G. T., Rahman, A., Humphrey, L. A., Bradley, T. J., Mueller, L. D., & Rose, M. R. (2023). Building bridges from genome to physiology using machine learning and *Drosophila* experimental evolution. *Physiological and Biochemical Zoology, 96*(3), 192–205. https://doi.org/10.1086/724827

Kezos, J. N., Cabral, L. G., Wong, B. D., Khou, B. K., Oh, A., Harb, J. F., Chiem, D., Bradley, T. J., Mueller, L. D., & Rose, M. R. (2017). Starvation but not locomotion enhances heart robustness in *Drosophila*. *Journal of Insect Physiology, 99*, 8–14. https://doi.org/10.1016/j.jinsphys.2017.03.004

Kezos, J. N., Phillips, M. A., Thomas, M. D., Ewunkem, A. J., Rutledge, G. A., Barter, T. T., Santos, M. A., Wong, B. D., Arnold, K. R., Humphrey, L. A., Yan, A., Nouzille, C., Sanchez, I., Cabral, L. G., Bradley, T. J., Mueller, L. D., Graves, J. L., & Rose, M. R. (2019). Genomics of early cardiac dysfunction and mortality in obese *Drosophila melanogaster*. *Physiological and Biochemical Zoology, 92*(6), 591–611. https://doi.org/10.1086/706099

Khan, A. I., Dinh, D. M., Schneider, D., Lenski, R. E., & Cooper, T. F. (2011). Negative epistasis between beneficial mutations in an evolving bacterial population. *Science, 332*(6034), 1193–1196. https://doi.org/10.1126/science.1203801

King, N. (2004). The unicellular ancestry of animal development. *Developmental Cell, 7*, 313–325. https://doi.org/10.1016/j.devcel.2004.08.010

Kofler, R., Gómez-Sánchez, D., & Schlötterer, C. (2016). PoPoolationTE2: Comparative population genomics of transposable elements using Pool-Seq. *Molecular Biology and Evolution, 33*(10), 2759–2764. https://doi.org/10.1093/molbev/msw137

Kofler, R., & Maloy, S. (2012). *Microbes and evolution: The world that Darwin never saw*. American Society for Microbiology Press. Washington, DC, USA.

Kofler, R., Orozco-terWengel, P., De Maio, N., Pandey, R. V., Nolte, V., Futschik, A., Kosiol, C., & Schlötterer, C. (2011). PoPoolation: A toolbox for population genetic analysis of next generation sequencing data from pooled individuals. *PLoS One, 6*(1), e15925. https://doi.org/10.1371/journal.pone.0015925

Kofler, R., Pandey, R. V., & Schlötterer, C. (2011). PoPoolation2: Identifying differentiation between populations using sequencing of pooled DNA samples (Pool-Seq). *Bioinformatics, 27*(24), 3435–3436. https://doi.org/10.1093/bioinformatics/btr589

Koga, H., Matsubara, M., Fujitani, H., Miyamoto, N., Komatsu, M., Kiyomoto, M., Akasaka, K., & Wada, H. (2010). Functional evolution of Ets in echinoderms with focus on the evolution of echinoderm larval skeletons. *Development Genes and Evolution, 220*(3–4), 107–115. https://doi.org/10.1007/s00427-010-0333-5

Kosuda, K. (1985). The aging effect on male mating activity in *Drosophila melanogaster*. *Behavior Genetics, 15*(3), 297–303. https://doi.org/10.1007/BF01065984

Kovács, Á. T., & Dragoš, A. (2019). Evolved biofilm: Review on the experimental evolution studies of *Bacillus subtilis* pellicles. *Journal of Molecular Biology, 431*(23), 4749–4759. https://doi.org/10.1016/j.jmb.2019.02.005

Krieber, M., & Rose, M. R. (1986). Males, parthenogenesis, and the maintenance of anisogamous sex. *Journal of Theoretical Biology, 122*(4), 421–440. https://doi.org/10.1016/S0022-5193(86)80183-7

Kubitschek, H. E. (1970). *Introduction to research with continuous cultures.* Prentice Hall. Englewood Cliffs, NJ, USA.

Kulski, J. K. (2016). Next-generation sequencing — An overview of the history, tools, and "omic" applications. In J. K. Kulski (Ed.), *Next generation sequencing — Advances, applications and challenges* (pp. 3–60). InTech, Rijeka, Croatia. https://doi.org/10.5772/61964

Lai, W. Y., Otte, K. A., & Schlötterer, C. (2023). Evolution of metabolome and transcriptome supports a hierarchical organization of adaptive traits. *Genome Biology and Evolution, 15*(6), evad098. https://doi.org/10.1093/gbe/evad098

Lake, D. M., Matson, Z. W. D., & Lenski, R. E. (2025). *STEPS User Manual* (v2.0.0). Zenodo.org. https://doi.org/10.5281/zenodo.16690374

Lang, G. I., Rice, D. P., Hickman, M. J., Sodergren, E., Weinstock, G. M., Botstein, D., & Desai, M. M. (2013). Pervasive genetic hitchhiking and clonal interference in forty evolving yeast populations. *Nature, 500,* 571–574. https://doi.org/10.1038/nature12344

Larsen, T. J., Jahan, I., Brock, D. A., Strassmann, J. E., & Queller, D. C. (2023). Reduced social function in experimentally evolved *Dictyostelium discoideum* implies selection for social conflict in nature. *Proceedings of the Royal Society B: Biological Sciences, 290*(2013), 20231722. https://doi.org/10.1098/rspb.2023.1722

Lasky, J. R., Josephs, E. B., & Morris, G. P. (2023, January 2). Genotype–environment associations to reveal the molecular basis of environmental adaptation. *The Plant Cell, 35*(1), 125–138. https://doi.org/10.1093/plcell/koac267

Le Gac, M., Plucain, J., Hindré, T., Lenski, R. E., & Schneider, D. (2012). Ecological and evolutionary dynamics of coexisting lineages during a long-term experiment with *Escherichia coli. Proceedings of the National Academy of Sciences of the United States of America, 109*(24), 9487–9492. https://doi.org/10.1073/pnas.1207091109

Leiby, N., & Marx, C. J. (2014). Metabolic erosion primarily through mutation accumulation, and not tradeoffs, drives limited evolution of substrate specificity in *Escherichia coli. PLoS Biology, 12*(2), e1001789. https://doi.org/10.1371/journal.pbio.1001789

Lenski, R. E. (1988). Experimental studies of pleiotropy and epistasis in *Escherichia coli*. II. Compensation for maladaptive pleiotropic effects associated with resistance to virus T4. *Evolution, 42*(3), 433–440. https://doi.org/10.1111/j.1558-5646.1988.tb04150.x

Lenski, R. E. (2017a). Convergence and divergence in a long-term experiment with bacteria. *The American Naturalist, 190*(Suppl. 1), S57–S68. https://doi.org/10.1086/691209

Lenski, R. E. (2017b). What is adaptation by natural selection? Perspectives of an experimental microbiologist. *PLOS Genetics, 13*(4), e1006668. https://doi.org/10.1371/journal.pgen.1006668

Lenski, R. E. (2017c). Experimental evolution and the dynamics of adaptation and genome evolution in microbial populations. *ISME Journal, 11*(10), 2181–2194. https://doi.org/10.1038/ismej.2017.69

Lenski, R. E. (2023). Revisiting the design of the long-term evolution experiment with *Escherichia coli*. *Journal of Molecular Evolution, 91*(3), 241–253. https://doi.org/10.1007/s00239-023-10095-3

Lenski, R. E., Rose, M. R., Simpson, S. C., & Tadler, S. C. (1991). Long-term experimental evolution in *Escherichia coli*. *I. Adaptation and divergence during 2,000 generations. The American Naturalist, 138*(6), 1315–1341. https://doi.org/10.1086/285289

Lenski, R. E., & Service, P. M. (1982). The statistical analysis of population growth rates calculated from schedules of survivorship and fecundity. *Ecology, 63*(3), 655–662. https://doi.org/10.2307/1936785

Lenski, R. E., & Travisano, M. (1994). Dynamics of adaptation and diversification: A 10,000-generation experiment with bacterial populations. *Proceedings of the National Academy of Sciences of the United States of America, 91*(15), 6808–6814. https://doi.org/10.1073/pnas.91.15.6808

Lenski, R. E., Wiser, M. J., Ribeck, N., Blount, Z. D., Nahum, J. R., Morris, J. J., Zaman, L., Turner, C. B., Wade, B. D., Maddamsetti, R., Burmeister, A. R., Baird, E. J., Bundy, J., Grant, N. A., Card, K. J., Rowles, M., Weatherspoon, K., Papoulis, S. E., Sullivan, R.,. . . . Hajela, N. (2015). Sustained fitness gains and variability in fitness trajectories in the long-term evolution experiment with *Escherichia coli*. *Proceedings of the Royal Society B: Biological Sciences, 282*(1821), 20152292. https://doi.org/10.1098/rspb.2015.2292

Leroi, A. M., Chen, W. R., & Rose, M. R. (1994). Long-term laboratory evolution of a genetic life-history trade-off in *Drosophila melanogaster*. 2. Stability of genetic correlations. *Evolution, 48*(4), 1258–1268. https://doi.org/10.1111/j.1558-5646.1994.tb05310.x

Leroi, A. M., Chippindale, A. K., & Rose, M. R. (1994). Long-term laboratory evolution of a genetic life-history trade-off in *Drosophila melanogaster*. 1. The role of genotype-by-environment interaction. *Evolution*, *48*(4), 1244–1257. https://doi.org/10.1111/j.1558-5646.1994.tb05309.x

Levins, R. (1966). The strategy of model building in population biology. *American Scientist*, *54*(4), 421–431. https://www.jstor.org/stable/27836590

Levy, S. B. (2002). *The antibiotic paradox: How the misuse of antibiotics destroys their curative powers* (2nd ed.). Perseus Publishing. Cambridge, MA, USA.

Lewontin, R. C. (1974). *The genetic basis of evolutionary change*. Columbia University Press. New York, USA.

Libby, E., Ratcliff, W., Travisano, M., & Kerr, N. (2014). Geometry shapes evolution of early multicellularity. *PLoS Computational Biology*, *10*(9), e1003803. https://doi.org/10.1371/journal.pcbi.1003803

Lind, P. A., Libby, E., Herzog, J., & Rainey, P. B. (2019). Predicting mutational routes to new adaptive phenotypes. *eLife*, *8*, e38822. https://doi.org/10.7554/eLife.38822

Linder, R. A., Zabanavar, B., Majumder, A., Hoang, H. C., Delgado, V. G., Tran, R., La, V. T., Leemans, S. W., & Long, A. D. (2022). Adaptation in outbred sexual yeast is repeatable, polygenic and favors rare haplotypes. *Molecular Biology and Evolution*, *39*(12), msac248. https://doi.org/10.1093/molbev/msac248

Linnen, C., Tatar, M., & Promislow, D. (2001). Cultural artifacts: A comparison of senescence in natural, laboratory-adapted and artificially selected lines of *Drosophila melanogaster*. *Evolutionary Ecology Research*, *3*(8), 877–888.

Long, H., Sung, W., Kucukyildirim, S., Williams, E., Miller, S. F., Guo, W., Patterson, C., Gregory, C., Strauss, C., Stone, C., Berne, C., Kysela, D., Shoemaker, W. R., Muscarella, M. E., Luo, H., Lennon, J. T., Brun, Y. V., & Lynch, M. (2018). Evolutionary determinants of genome-wide nucleotide composition. *Nature Ecology & Evolution*, *2*(2), 237–240. https://doi.org/10.1038/s41559-017-0425-y

Losos, J. B. (2009). *Lizards in an evolutionary tree: Ecology and adaptive radiation of anoles*. University of California Press. Oakland, CA, USA.

Losos, J. B. (2018). *Improbable destinies: Fate, chance, and the future of evolution*. Riverhead Books. New York, NY, USA.

Luckinbill, L. S., Arking, R., Clare, M. J., Cirocco, W. C., & Buck, S. A. (1984). Selection for delayed senescence in *Drosophila melanogaster*. *Evolution*, *38*(5), 996–1003. https://doi.org/10.1111/j.1558-5646.1984.tb00369.x

Maddamsetti, R., Hatcher, P. J., Green, A. G., Williams, B. L., Marks, D. S., & Lenski, R. E. (2017). Core genes evolve rapidly in the long-term evolution experiment with *Escherichia coli*. *Genome Biology and Evolution, 9*(4), 1072–1083. https://doi.org/10.1093/gbe/evx064

Maddamsetti, R., & Lenski, R. E. (2018). Analysis of bacterial genomes from an evolution experiment with horizontal gene transfer shows that recombination can sometimes overwhelm selection. *PLOS Genetics, 14*(1), e1007199. https://doi.org/10.1371/journal.pgen.1007199

Maddamsetti, R., Lenski, R. E., & Barrick, J. E. (2015). Adaptation, clonal interference, and frequency-dependent interactions in a long-term evolution experiment with *Escherichia coli*. *Genetics, 200*(3), 619–631. https://doi.org/10.1534/genetics.115.176677

Magalhães, S., Blanchet, E., Egas, M., & Olivieri, I. (2011). Environmental effects on the detection of adaptation. *Journal of Evolutionary Biology, 24*(12), 2653–2662. https://doi.org/10.1111/j.1420-9101.2011.02388.x

Manhes, P., & Velicer, G. J. (2011). Experimental evolution of selfish policing in social bacteria. *Proceedings of the National Academy of Sciences of the United States of America, 108*(20), 8357–8362. https://doi.org/10.1073/pnas.1014695108

Mardis, E. R. (2011). A decade's perspective on DNA sequencing technology. *Nature, 470*(7333), 198–203. https://doi.org/10.1038/nature09796

Marek, A., & Korona, R. (2016). Strong dominance of functional alleles over gene deletions in both intensely growing and deeply starved yeast cells. *Journal of Evolutionary Biology, 29*(9), 1836–1845. https://doi.org/10.1111/jeb.12917

Markov, P. V., Ghafari, M., Beer, M., Lythgoe, K., Simmonds, P., Stilianakis, N. I., & Katzourakis, A. (2023). The evolution of SARS-CoV-2. *Nature Reviews Microbiology, 21*(6), 361–379. https://doi.org/10.1038/s41579-023-00878-2

Marshall, D. J., Malerba, M., Lines, T., Sezmis, A. L., Hasan, C. M., Lenski, R. E., & McDonald, M. J. (2022). Long-term experimental evolution decouples size and production costs in *Escherichia coli*. *Proceedings of the National Academy of Sciences of the United States of America, 119*(21), e2200713119. https://doi.org/10.1073/pnas.2200713119

Martínez, D. E. (1998). Mortality patterns suggest lack of senescence in *Hydra*. *Experimental Gerontology, 33*(3), 217–225. https://doi.org/10.1016/S0531-5565(97)00113-7

Matos, M., Avelar, T., & Rose, M. R. (2002). Variation in the rate of convergent evolution: Adaptation to a laboratory environment in *Drosophila subobscura. Journal of Evolutionary Biology, 15*(4), 673–682. https://doi. org/10.1046/j.1420-9101.2002.00405.x

Matos, M., Rego, C., Levy, A., Teotónio, H., & Rose, M. R. (2000a). An evolutionary no man's land. *Trends in Ecology & Evolution, 15*(5), 206. https://doi.org/10.1016/S0169-5347(00)01844-9

Matos, M., Rose, M. R., Rocha Pité, M. T., Rego, C., & Avelar, T. (2000b). Adaptation to the laboratory environment in *Drosophila subobscura. Journal of Evolutionary Biology, 13*(1), 9–19. https://doi.org/10.1046/j. 1420-9101.2000.00116.x

Maynard Smith, J. (1958). The effects of temperature and of egg-laying on the longevity of *Drosophila subobscura. Journal of Experimental Biology, 35*(4), 832–842. https://doi.org/10.1242/jeb.35.4.832

Maynard Smith, J. (1978). *The evolution of sex.* Cambridge University Press. Cambridge, UK.

Mayr, E. (1963). *Animal species and evolution.* Belknap Press, Harvard University Press. Cambridge, MA, USA; London, UK.

Mayr, E. (1982). *The growth of biological thought: Diversity, evolution, and inheritance* (pp. 681–687). Harvard University Press. Cambridge, MA, USA.

McDonald, M. J., Gehrig, S. M., Meintjes, P. J., Zhang, X. X., & Rainey, P. B. (2009). Adaptive divergence in experimental populations of *Pseudomonas fluorescens.* IV. Genetic constraints guide evolutionary trajectories in a parallel adaptive radiation. *Genetics, 183*(3), 1041–1053. https://doi. org/10.1534/genetics.109.107110

McDonald, M. J.., Rice, D. P., & Desai, M. M. (2016). Sex speeds adaptation by altering the dynamics of molecular evolution. *Nature, 531*(7593), 233–236. https://doi.org/10.1038/nature17143

Medawar, P. B. (1952). *An unsolved problem of biology.* H. K. Lewis. London, UK.

Meyer, J. R., Agrawal, A. A., Quick, R. T., Dobias, D. T., Schneider, D., & Lenski, R. E. (2010). Parallel changes in host resistance to viral infection during 45,000 generations of relaxed selection. *Evolution, 64*(10), 3024–3034. https://doi.org/10.1111/j.1558-5646.2010.01049.x

Meyer, J. R., Dobias, D. T., Medina, S. J., Servilio, L., Gupta, A., & Lenski, R. E. (2016). Ecological speciation of bacteriophage lambda in allopatry and sympatry. *Science, 354*(6317), 1301–1304. https://doi.org/10.1126/science.aai8446

Meyer, J. R., Dobias, D. T., Weitz, J. S., Barrick, J. E., Quick, R. T., & Lenski, R. E. (2012). Repeatability and contingency in the evolution of a key innovation in phage lambda. *Science, 335*(6067), 428–432. https://doi.org/10.1126/ science.1214449

Michod, R. E., & Levin, B. R. (Eds.). (1988). *The evolution of sex*. Sinauer Associates. Sunderland, MA, USA.

Miller, S. L. (1953). A production of amino acids under possible primitive earth conditions. *Science, 117*(3046), 528–529. https://doi.org/10.1126/science.117.3046.528

Millicent, E., & Thoday, J. M. (1960). Gene flow and divergence under disruptive selection. *Science, 131*(3409), 1311–1312. https://doi.org/10.1126/science.131.3409.1311

Mueller, L. D. (1987). Evolution of accelerated senescence in laboratory populations of *Drosophila*. *Proceedings of the National Academy of Sciences of the United States of America, 84*(7), 1974–1977. https://doi.org/10.1073/pnas.84.7.1974

Mueller, L. D., Graves, J. L., & Rose, M. R. (1993). Interactions between density-dependent and age-specific selection in *Drosophila melanogaster*. *Functional Ecology, 7*, 469–479. https://doi.org/10.2307/2390034

Mueller, L. D., Joshi, A., Santos, M., & Rose, M. R. (2013). Effective population size and evolutionary dynamics in outbred laboratory populations of *Drosophila*. *Journal of Genetics, 92*(3), 349–361. https://doi.org/10.1007/s12041-013-0296-1

Mueller, L. D., Phillips, M. A., Barter, T. T., Greenspan, Z. S., & Rose, M. R. (2018). Genome-wide mapping of gene–phenotype relationships in experimentally evolved populations. *Molecular Biology and Evolution, 35*(8), 2085–2095. https://doi.org/10.1093/molbev/msy113

Mueller, L. D., Rauser, C. L., & Rose, M. R. (2011). *Does aging stop?* Oxford University Press. New York, NY, USA.

Mueller, L. D., & Rose, M. R. (1996). Evolutionary theory predicts late-life mortality plateaus. *Proceedings of the National Academy of Sciences, 93*(26), 15249–15253. https://doi.org/10.1073/pnas.93.26.15249

Muller, H. J. (1932). Some genetic aspects of sex. *The American Naturalist, 66*(703), 118–138. https://www.journals.uchicago.edu/doi/10.1086/280418

Muller, H. J. (1964). The relation of recombination to mutational advance. *Mutation Research, 1*(1), 2–9. https://doi.org/10.1016/0027-5107(64)90047-8

Naas, T., Blot, M., Fitch, W. M., & Arber, W. (1995). Dynamics of IS-related genetic rearrangements in resting *Escherichia coli* K-12. *Molecular Biology and Evolution, 12*(2), 198–207. https://doi.org/10.1093/oxfordjournals.molbev.a040198

Nagylaki, T. (1977). The evolution of one- and two-locus systems. II. *Genetics, 85*(2), 347–354. https://doi.org/10.1093/genetics/85.2.347

Nghiem, D., Gibbs, A. G., Rose, M. R., & Bradley, T. J. (2000). Postponed aging and desiccation resistance in *Drosophila melanogaster*. *Experimental Gerontology*, *35*(8), 957–969. https://doi.org/10.1016/S0531-5565(00)00163-7

Noble, L. M., Chelo, I., Guzella, T., Afonso, B., Riccardi, D. D., Ammerman, P., Dayarian, A., Carvalho, S., Crist, A., Pino-Querido, A., Shraiman, B., Rockman, M. V., & Teotónio, H. (2017). Polygenicity and epistasis underlie fitness-proximal traits in the *Caenorhabditis elegans* multiparental experimental evolution (CeMEE) panel. *Genetics*, *207*(4), 1663–1685. https://doi.org/10.1534/genetics.117.300406

Nouhaud, P., Tobler, R., Nolte, V., & Schlötterer, C. (2016). Ancestral population reconstitution from isofemale lines as a tool for experimental evolution. *Ecology and Evolution*, *6*(20), 7169–7175. https://doi.org/10.1002/ece3.2402

Nuzhdin, S. V., & Turner, T. L. (2013). Promises and limitations of hitchhiking mapping. *Current Opinion in Genetics & Development*, *23*(6), 694–699. https://doi.org/10.1016/j.gde.2013.10.002

Orgogozo, V. (2015). Replaying the tape of life in the twenty-first century. *Interface Focus*, *5*(6), 20150057. https://doi.org/10.1098/rsfs.2015.0057

Orozco-Terwengel, P., Kapun, M., Nolte, V., Kofler, R., Flatt, T., & Schlötterer, C. (2012). Adaptation of *Drosophila* to a novel laboratory environment reveals temporally heterogeneous trajectories of selected alleles. *Molecular Ecology*, *21*(20), 4931–4941. https://doi.org/10.1111/j.1365-294X.2012.05673.x

Passananti, H. B., Beckman, K. A., & Rose, M. R. (2004). Relaxed stress selection in *Drosophila melanogaster*. In M. R. Rose, H. B. Passananti, & M. Matos (Eds.), *Methuselah flies: A case study in the evolution of aging* (pp. 296–322). World Scientific Publishing. Singapore. https://doi.org/10.1142/9789812567222_0028

Passananti, H. B., Deckert-Cruz, D. J., Chippindale, A. K., Le, B. H., & Rose, M. R. (2004). Reverse evolution of aging in *Drosophila melanogaster*. In M. R. Rose, H. B. Passananti, & M. Matos (Eds.), *Methuselah flies: A case study in the evolution of aging* (pp. 323–352). World Scientific Publishing. Singapore. https://doi.org/10.1142/9789812567222_0027

Peng, F., Widmann, S., Wünsche, A., Duan, K., Donovan, K. A., Dobson, R. C. J., Lenski, R. E., & Cooper, T. F. (2018). Effects of beneficial mutations in *pykF* gene vary over time and across replicate populations in a long-term experiment with bacteria. *Molecular Biology and Evolution*, *35*(1), 202–210. https://doi.org/10.1093/molbev/msx279

Pennell, M. W., Harmon, L. J., & Uyeda, J. C. (2014). Is there room for punctuated equilibrium in macroevolution? *Trends in Ecology & Evolution*, *29*(1), 23–32. https://doi.org/10.1016/j.tree.2013.07.004

Petrie, K. L., Palmer, N. D., Johnson, D. T., Medina, S. J., Yan, S. J., Li, V., Burmeister, A. R., & Meyer, J. R. (2018). Destabilizing mutations encode nongenetic variation that drives evolutionary innovation. *Science*, *359*(6383), 1542–1545. https://doi.org/10.1126/science.aar1954

Phelan, J. P., Archer, M. A., Beckman, K. A., Chippindale, A. K., Nusbaum, T. J., & Rose, M. R. (2003). Breakdown in correlations during laboratory evolution. I. Comparative analyses of *Drosophila* populations. *Evolution*, *57*(3), 527–535. https://doi.org/10.1111/j.0014-3820.2003.tb01544.x

Phillips, M. A., Arnold, K. R., Vue, Z., Beasley, H. K., Garza-Lopez, E., Marshall, A. G., Morton, D. J., McReynolds, M. R., Barter, T. T., & Hinton, A. (2022). Combining metabolomics and experimental evolution reveals key mechanisms underlying longevity differences in laboratory evolved *Drosophila melanogaster* populations. *International Journal of Molecular Sciences*, *23*(3), 1067. https://doi.org/10.3390/ijms23031067

Phillips, M. A., Briar, R. K., Scaffo, M., Zhou, S., & Burke, M. K. (2022). Strength of selection potentiates distinct adaptive responses in an evolution experiment with outcrossing yeast. *bioRxiv*, 2022.05.19.492575. https://doi.org/10.1101/2022.05.19.492575

Phillips, M. A., Long, A. D., Greenspan, Z. S., Greer, L. F., Burke, M. K., Villeponteau, B., Matsagas, K. C., Rizza, C. L., Mueller, L. D., & Rose, M. R. (2016). Genome-wide analysis of long-term evolutionary domestication in *Drosophila melanogaster*. *Scientific Reports*, *6*(1), Article 39281. https://doi.org/10.1038/srep39281

Phillips, M. A., Rutledge, G. A., Kezos, J. N., Greenspan, Z. S., Talbott, A., Matty, S., Arain, H., Mueller, L. D., Rose, M. R., & Shahrestani, P. (2018). Effects of evolutionary history on genome-wide and phenotypic convergence in *Drosophila* populations. *BMC Genomics*, *19*(1), Article 261. https://doi.org/10.1186/s12864-018-5118-7

Poltak, S. R., & Cooper, V. S. (2011). Ecological succession in long-term experimentally evolved biofilms produces synergistic communities. *ISME Journal*, *5*(3), 369–378. https://doi.org/10.1038/ismej.2010.136

Quan, S., Ray, J. C., Kwota, Z., Duong, T., Balázsi, G., Cooper, T. F., & Monds, R. D. (2012). Adaptive evolution of the lactose utilization network in experimentally evolved populations of *Escherichia coli*. *PLoS Genetics*, *8*(1), e1002444. https://doi.org/10.1371/journal.pgen.1002444

Quandt, E. M., Deatherage, D. E., Ellington, A. D., Georgiou, G., & Barrick, J. E. (2014). Recursive genomewide recombination and sequencing reveals a key refinement step in the evolution of a metabolic innovation in *Escherichia coli*. *Proceedings of the National Academy of Sciences of the United States of America*, *111*(6), 2217–2222. https://doi.org/10.1073/pnas.1314561111

Quandt, E. M., Gollihar, J., Blount, Z. D., Ellington, A. D., Georgiou, G., & Barrick, J. E. (2015). Fine-tuning citrate synthase flux potentiates and refines metabolic innovation in the Lenski evolution experiment. *eLife*, *4*, e09696. https://doi.org/10.7554/eLife.09696

Rainey, P. B., & Travisano, M. (1998). Adaptive radiation in a heterogeneous environment. *Nature*, *394*(6688), 69–72. https://doi.org/10.1038/27900

Rang, C. U., Peng, A. Y., & Chao, L. (2011). Temporal dynamics of bacterial aging and rejuvenation. *Current Biology*, *21*(21), 1813–1816. https://doi.org/10.1016/j.cub.2011.09.018

Ratcliff, W. C., Denison, R. F., Borrello, M., & Travisano, M. (2012). Experimental evolution of multicellularity. *Proceedings of the National Academy of Sciences of the United States of America*, *109*(5), 1595–1600. https://doi.org/10.1073/pnas.1115323109

Ratcliff, W. C., Fankhauser, J. D., Rogers, D. W., Greig, D., & Travisano, M. (2015). Origins of multicellular evolvability in snowflake yeast. *Nature Communications*, *6*, Article 6102. https://doi.org/10.1038/ncomms7102

Rauser, C. L., Abdel-Aal, Y., Shieh, J. A., Suen, C. W., Mueller, L. D., & Rose, M. R. (2005). Lifelong heterogeneity in fecundity is insufficient to explain late-life fecundity plateaus in *Drosophila melanogaster*. *Experimental Gerontology*, *40*(8–9), 660–670. https://doi.org/10.1016/j.exger.2005.06.006

Rauser, C. L., Hong, J. S., Cung, M. B., Pham, K. M., Mueller, L. D., & Rose, M. R. (2005). Testing whether male age or high nutrition causes the cessation of reproductive aging in female *Drosophila melanogaster* populations. *Rejuvenation Research*, *8*(2), 86–95. https://doi.org/10.1089/rej.2005.8.86

Rauser, C. L., Mueller, L. D., & Rose, M. R. (2003). Aging, fertility, and immortality. *Experimental Gerontology*, *38*(1–2), 27–33. https://doi.org/10.1016/S0531-5565(02)00148-1

Rauser, C. L., Mueller, L. D., Travisano, M., & Rose, M. R. (2009). Evolution of aging and late life. In T. Garland & M. R. Rose (Eds.), *Experimental evolution: Concepts, methods, and applications of selection* (pp. 551–584). University of California. Berkeley, USA.

Rauser, C. L., Tierney, J. J., Gunion, S. M., Covarrubias, G. M., Mueller, L. D., & Rose, M. R. (2006). Evolution of late-life fecundity in *Drosophila melanogaster*. *Journal of Evolutionary Biology*, *19*(1), 289–301. https://doi.org/10.1111/j.1420-9101.2005.00966.x

Remolina, S. C., Chang, P. L., Leips, J., Nuzhdin, S. V., & Hughes, K. A. (2012). Genomic basis of aging and life-history evolution in *Drosophila melanogaster*. *Evolution*, *66*(11), 3390–3403. https://doi.org/10.1111/j.1558-5646. 2012.01710.x

Rendueles, O., Zee, P. C., Dinkelacker, I., Amherd, M., Wielgoss, S., & Velicer, G. J. (2015). Rapid and widespread de novo evolution of kin discrimination. *Proceedings of the National Academy of Sciences*, *112*(29), 9076–9081. https://doi.org/10.1073/pnas.1502251112

Rezende, E., Balanyà, J., Rodríguez-Trelles, F., Rego, C., Fragata, I., Matos, M., Serra, L., & Santos, M. (2010). Climate change and chromosomal inversions in *Drosophila subobscura*. *Climate Research*, *43*(1), 103–114. https://doi.org/10.3354/cr00869

Robinson, C. E., Thyagarajan, H., & Chippindale, A. K. (2023). Evolution of reproductive isolation in a long-term evolution experiment with *Drosophila melanogaster*: 30 years of divergent life-history selection. *Evolution*, *77*(8), 1756–1768. https://doi.org/10.1093/evolut/qpad098

Rode, N. O., Holtz, Y., Loridon, K., Santoni, S., Ronfort, J., & Gay, L. (2018). How to optimize the precision of allele and haplotype frequency estimates using pooled-sequencing data. *Molecular Ecology Resources*, *18*(2), 194–203. https://doi.org/10.1111/1755-0998.12723

Rose, M. R. (1983). The contagion mechanism for the origin of sex. *Journal of Theoretical Biology*, *101*(1), 137–146. https://doi.org/10.1016/0022-5193 (83)90277-1

Rose, M. R. (1984a). Genetic covariation in *Drosophila* life history: Untangling the data. *The American Naturalist*, *123*(4), 565–569. https://doi.org/10.1086/284222

Rose, M. R. (1984b). Laboratory evolution of postponed senescence in *Drosophila melanogaster*. *Evolution*, *38*(5), 1004–1010. https://doi.org/10.1111/j.1558-5646.1984.tb00370.x

Rose, M. R. (1985). Life-history evolution with antagonistic pleiotropy and overlapping generations. *Theoretical Population Biology*, *28*(3), 342–358. https://doi.org/10.1016/0040-5809(85)90034-6

Rose, M. R. (1991). *Evolutionary biology of aging*. Oxford University Press. New York, NY, USA.

Rose, M. R. (2005). *The long tomorrow: How advances in evolutionary biology can help us postpone aging*. Oxford University Press. New York, USA.

Rose, M. R., & Charlesworth, B. (1980). A test of evolutionary theories of senescence. *Nature*, *287*, 141–142. https://doi.org/10.1038/287141a0

Rose, M. R., & Charlesworth, B. (1981a). Genetics of life history in *Drosophila melanogaster*. I. Sib analysis of adult females. *Genetics*, *97*(1), 173–186. https://doi.org/10.1093/genetics/97.1.173

Rose, M. R., & Charlesworth, B. (1981b). Genetics of life history in *Drosophila melanogaster*. II. Exploratory selection experiments. *Genetics*, *97*(1), 187–196. https://doi.org/10.1093/genetics/97.1.187

Rose, M. R., & Doolittle, W. F. (1983). Molecular biological mechanisms of speciation. *Science*, *220*(4593), 157–162. https://doi.org/10.1126/science.220.4593.157

Rose, M. R., Drapeau, M. D., Yazdi, P. G., Shah, K. H., Moise, D. B., Thakar, R. R., Rauser, C. L., & Mueller, L. D. (2002). Evolution of late-life mortality in *Drosophila melanogaster*. *Evolution*, *56*(10), 1982–1991. https://doi.org/10.1111/j.0014-3820.2002.tb00124.x

Rose, M. R., Graves, J. L., & Hutchinson, E. W. (1990). The use of selection to probe patterns of pleiotropy in fitness characters. In F. Gilbert (Ed.), *Insect life cycles: Genetics, evolution and coordination* (pp. 29–42). Springer-Verlag. Berlin. https://doi.org/10.1007/978-1-4471-3464-0_3

Rose, M. R., & Lauder, G. V. (Eds.). (1996). *Adaptation*. Academic Press. New York, USA.

Rose, M. R., Nusbaum, T. J., & Chippindale, A. K. (1996). Laboratory evolution: The experimental wonderland and the Cheshire Cat syndrome. In M. R. Rose & G. V. Lauder (Eds.), *Adaptation* (pp. 221–241). Academic Press. New York. USA.

Rose, M. R., Passananti, H. B., & Matos, M. (Eds.) (2004). *Methuselah flies: A case study in the evolution of aging*. World Scientific Publishing. Singapore. https://doi.org/10.1142/5457

Rose, M. R., Vu, L. N., Park, S. U., & Graves, J. L. (1992). Selection on stress resistance increases longevity in *Drosophila melanogaster*. *Experimental Gerontology*, *27*(2), 241–250. https://doi.org/10.1016/0531-5565(92)90048-5

Rozen, D. E., & Lenski, R. E. (2000). Long-term experimental evolution in *Escherichia coli* VIII. Dynamics of a balanced polymorphism. *The American Naturalist*, *155*(1), 24–35. https://doi.org/10.1086/303299

Rozen, D. E., Schneider, D., & Lenski, R. E. (2005). Long-term experimental evolution in *Escherichia coli* XIII. Phylogenetic history of a balanced polymorphism. *Journal of Molecular Evolution*, *61*(2), 171–180. https://doi.org/10.1007/s00239-004-0322-2

Rudman, S. M., Greenblum, S. I., Rajpurohit, S., Betancourt, N. J., Hanna, J., Tilk, S., Yokoyama, T., Petrov, D. A., & Schmidt, P. (2022). Direct observation of adaptive tracking on ecological time scales in *Drosophila*. *Science*, *375*(6586), eabj7484. https://doi.org/10.1126/science.abj7484

Rundle, H. D., Chenoweth, S. F., Doughty, P., & Blows, M. W. (2005). Divergent selection and the evolution of signal traits and mating preferences. *PLoS Biology*, *3*(11), e368. https://doi.org/10.1371/journal.pbio.0030368

Rutledge, G. A., Cabral, L. G., Kuey, B. J., Lee, J. D., Mueller, L. D., & Rose, M. R. (2020). Hamiltonian patterns of age-dependent adaptation to novel environments. *PLoS ONE, 15*(10), e0240132. https://doi.org/10.1371/journal.pone.0240132

Santos, J., Pascual, M., Simões, P., Fragata, I., Lima, M., Kellen, B., Santos, M., Marques, A., Rose, M. R., & Matos, M. (2012). From nature to the laboratory: The impact of founder effects on adaptation. *Journal of Evolutionary Biology, 25*(12), 2607–2622. https://doi.org/10.1111/jeb.12008

Santos, J., Pascual, M., Simões, P., Fragata, I., Rose, M. R., & Matos, M. (2013). Fast evolutionary genetic differentiation during experimental colonizations. *Journal of Genetics, 92*(2), 183–194. https://doi.org/10.1007/s12041-013-0239-x

Santos, M. A., Antunes, M. A., Grandela, A., Quina, A. S., Santos, M., Matos, M., & Simões, P. (2023). Slow and population-specific evolutionary response to a warming environment. *Scientific Reports, 13*, Article 9700. https://doi.org/10.1038/s41598-023-36273-3

Santos, M. A., Carromeu-Santos, A., Quina, A. S., Antunes, M. A., Kristensen, T. N., Santos, M., Matos, M., Fragata, I., & Simões, P. (2024). Experimental Evolution in a Warming World: The Omics Era. *Molecular Biology and Evolution, 41*(8), msae148. https://10.1093/molbev/msae148

Santos, M., Brites, D., & Laayouni, H. (2006). Thermal evolution of pre-adult life history traits, geometric size and shape, and developmental stability in *Drosophila subobscura*. *Journal of Evolutionary Biology, 19*(6), 2006–2021. https://doi.org/10.1038/s41598-023-36273-3

Scheutz, F., & Strockbine, N. A. (2005). Genus I. *Escherichia* Castellani and Chalmers 1919. In G. M. Garrity, D. J. Brenner, N. R. Krieg, J. R. Staley, & G. M. Garrity (Eds.), *Bergey's manual of systematic bacteriology: Volume 2. The proteobacteria, Part B: The gammaproteobacteria* (2nd ed., pp. 607–624). Springer. New York, NY, USA.

Schlötterer, C. (2023). How predictable is adaptation from standing genetic variation? Experimental evolution in *Drosophila* highlights the central role of redundancy and linkage disequilibrium. *Philosophical Transactions of the Royal Society B, 378*(1877), 20220046. https://doi.org/10.1098/rstb.2022.0046

Schlötterer, C., Kofler, R., Versace, E., Tobler, R., & Franssen, S. U. (2015). Combining experimental evolution with next-generation sequencing: A powerful tool to study adaptation from standing genetic variation. *Heredity, 114*, 431–440. https://doi.org/10.1038/hdy.2014.86

Seabra, S. G., Fragata, I., Antunes, M. A., Faria, G. S., Santos, M. A., Sousa, V. C., Simões, P., & Matos, M. (2018). Different genomic changes underlie adaptive evolution in populations of contrasting history. *Molecular Biology and Evolution, 35*(3), 549–563. https://doi.org/10.1093/molbev/msx247

Seger, J., & Hamilton, W. D. (1988). Parasites and sex. In R. E. Michod & B. R. Levin (Eds.), *The evolution of sex* (pp. 176–193). Sinauer Associates. Sunderland, MA, USA.

Service, P. M. (1987). Physiological mechanisms of increased stress resistance in *Drosophila melanogaster* selected for postponed senescence. *Physiological Zoology, 60*(3), 321–326. https://doi.org/10.1086/physzool.60.3.30162285

Service, P. M. (1989). The effect of mating status on lifespan, egg laying, and starvation resistance in *Drosophila melanogaster* in relation to selection on longevity. *Journal of Insect Physiology, 35*(5), 447–452. https://doi.org/10.1016/0022-1910(89)90120-0

Service, P. M., Hutchinson, E. W., MacKinley, M. D., & Rose, M. R. (1985). Resistance to environmental stress in *Drosophila melanogaster* selected for postponed senescence. *Physiological Zoology, 58*(4), 380–389. https://doi.org/10.1086/physzool.58.4.30156013

Service, P. M., Hutchinson, E. W., & Rose, M. R. (1988). Multiple genetic mechanisms for the evolution of senescence in *Drosophila melanogaster*. *Evolution, 42*(4), 708–715. https://doi.org/10.2307/2408862

Service, P. M., & Rose, M. R. (1985). Genetic covariation among life-history components: The effect of novel environments. *Evolution, 39*(4), 943–951. https://doi.org/10.2307/2408694

Sgrò, C. M., & Partridge, L. (2000). Evolutionary responses of the life history of wild-caught *Drosophila melanogaster* to two standard methods of laboratory culture. *The American Naturalist, 156*(4), 341–353. https://doi.org/10.1086/303394

Shahrestani, P., Burke, M. K., Birse, R., Kezos, J. N., Ocorr, K., Mueller, L. D., Rose, M. R., & Bodmer, R. (2017). Experimental evolution and heart function in *Drosophila*. *Physiological and Biochemical Zoology, 90*(2), 281–293. https://doi.org/10.1086/689288

Shahrestani, P., Quach, J., Mueller, L. D., & Rose, M. R. (2012). Paradoxical physiological transitions from aging to late life in *Drosophila*. *Rejuvenation Research, 15*(1), 49–58. https://doi.org/10.1089/rej.2011.1201

Shapiro, B. J., Friedman, J., Cordero, O. X., Preheim, S. P., Timberlake, S. C., Szabó, G., Polz, M. F., & Alm, E. J. (2012). Population genomics of early events in the ecological differentiation of bacteria. *Science*, *336*(6077), 48–51. https://doi.org/10.1126/science.1218198

Simões, P., Fragata, I., Santos, J., Santos, M. A., Santos, M., Rose, M. R., & Matos, M. (2019). How phenotypic convergence arises in experimental evolution. *Evolution*, *73*(9), 1839–1849. https://doi.org/10.1111/evo.13806

Simões, P., Fragata, I., Seabra, S. G., Faria, G. S., Santos, M. A., Rose, M. R., Santos, M., & Matos, M. (2017). Predictable phenotypic, but not karyotypic, evolution of populations with contrasting initial history. *Scientific Reports*, *7*(1), Article 913. https://doi.org/10.1038/s41598-017-00968-1

Simões, P., Rose, M. R., Duarte, A., Gonçalves, R., & Matos, M. (2007). Evolutionary domestication in *Drosophila subobscura*. *Journal of Evolutionary Biology*, *20*(2), 758–766. https://doi.org/10.1111/j.1420-9101.2006.01244.x

Simões, P., Santos, J., Fragata, I., Mueller, L. D., Rose, M. R., & Matos, M. (2008). How repeatable is adaptive evolution? The role of geographical origin and founder effects in laboratory adaptation. *Evolution*, *62*(8), 1817–1829. https://doi.org/10.1111/j.1558-5646.2008.00423.x

Simões, P., Santos, J., & Matos, M. (2009). Experimental evolutionary domestication. In T. Garland & M. R. Rose (Eds.), *Experimental evolution: Concepts, methods, and applications of selection experiments* (pp. 89–110). University of California Press. Berkeley, CA. USA. https://doi.org/10.1525/california/9780520247666.003.0005

Sniegowski, P. D., Gerrish, P. J., & Lenski, R. E. (1997). Evolution of high mutation rates in experimental populations of *Escherichia coli*. *Nature*, *387*(6634), 703–705. https://doi.org/10.1038/42701

Sokal, R. R. (1970). Senescence and genetic load: Evidence from *Tribolium*. *Science*, *167*(3926), 1733–1734. https://doi.org/10.1126/science.167.3926.1733

Souza, V., Turner, P. E., & Lenski, R. E. (1997). Long-term experimental evolution in *Escherichia coli* V. Effects of recombination with immigrant genotypes on the rate of bacterial evolution. *Journal of Evolutionary Biology*, *10*(5), 743–769. https://doi.org/10.1046/j.1420-9101.1997.10050743.x

Spealman, P., De, T, Chuong, J. N., & Gresham, D. (2023). Best practices in microbial experimental evolution: Using reporters and long-read sequencing to identify copy number variation in experimental evolution. *Journal of Molecular Evolution*, *91*(3), 356–368. https://doi.org/10.1007/s00239-023-10102-7

Spiers, A. J., Kahn, S. G., Bohannon, J., Travisano, M., & Rainey, P. B. (2002). Adaptive divergence in experimental populations of *Pseudomonas fluorescens*. I. Genetic and phenotypic bases of wrinkly spreader fitness. *Genetics, 161*(1), 33–46. https://doi.org/10.1093/genetics/161.1.33

Stearns, S. C. (1976). Life-history tactics: a review of the ideas. *Quarterly Review of Biology, 51*(1), 3–47. https://doi.org/10.1086/409052

Stern, D. L. (2013). The genetic causes of convergent evolution. *Nature Reviews Genetics, 14*, 751–764. https://doi.org/10.1038/nrg3483

Strassmann, J. E., Zhu, Y., & Queller, D. C. (2000). Altruism and social cheating in the social amoeba *Dictyostelium discoideum*. *Nature, 408*(6815), 965–967. https://doi.org/10.1038/35050087

Strobel, H. M., Sweetzel Labador, D., Basu, D., Sane, M., Corbett, K. D., & Meyer, J. R. (2024). Viral receptor-binding protein evolves new function through mutations that cause trimer instability and functional heterogeneity. *Molecular Biology and Evolution, 41*(4), msae056. https://doi.org/10.1093/molbev/msae056

Swallow, J. G., Carter, P. A., & Garland, T. (1998). Artificial selection for increased wheel-running behavior in house mice. *Behavior Genetics, 28*(3), 227–237. https://doi.org/10.1023/a:1021479331779

Swallow, J. G., Hayes, J. P., Koteja, P., & Garland, T. (2009). Selection experiments and experimental evolution of performance and physiology. In T. D. Garland & M. R. Rose (Eds.), *Experimental evolution: Concepts, methods, and applications of selection experiments* (pp. 301–351). University of California Press. Berkeley, CA, USA. https://doi.org/10.1525/california/9780520247666.003.0012

Szalay, T., & Golovchenko, J. A. (2015). De novo sequencing and variant calling with nanopores using PoreSeq. *Nature Biotechnology, 33*(10), 1087–1091. https://doi.org/10.1038/nbt.3360

Tajkarimi, M., Rhinehardt, K., Thomas, M., Ewunkem, J. A., Campbell, A., Boyd, S., Turner, D., Harrison, S. H., & Graves, J. L. Jr. (2017). Selection for ionic-confers silver nanoparticle resistance in *Escherichia coli*. *JSM Nanotechnology and Nanomedicine, 5*(1), 1047.

Tenaillon, O., Barrick, J. E., Ribeck, N., Deatherage, D. E., Blanchard, J. L., Dasgupta, A., Wu, G. C., Wielgoss, S., Cruveiller, S., Médigue, C., Schneider, D., & Lenski, R. E. (2016). Tempo and mode of genome evolution in a 50,000-generation experiment. *Nature, 536*(7615), 165–170. https://doi.org/10.1038/nature18959

Tenaillon, O., Rodríguez-Verdugo, A., Gaut, R. L., McDonald, P., Bennett, A. F., Long, A. D., & Gaut, B. S. (2012). The molecular diversity of adaptive convergence. *Science, 335*(6067), 457–461. https://doi.org/10.1126/science.1212986

Teotónio, H., Estes, S., Phillips, P. C., & Baer, C. F. (2017). Experimental evolution with *Caenorhabditis* nematodes. *Genetics, 206*(2), 691–716. https://doi.org/10.1534/genetics.115.186288

Teotónio, H., Matos, M., & Rose, M. R. (2002). Reverse evolution of fitness in *Drosophila melanogaster. Journal of Evolutionary Biology, 15*(4), 608–617. https://doi.org/10.1046/j.1420-9101.2002.00424.x

Teotónio, H., & Rose, M. R. (2000). Variation in the reversibility of evolution. *Nature, 408*(6811), 463–466. https://doi.org/10.1038/35044070

Thomas, M. D., Ewunkem, J. A., Boyd, S., Williams, D. K., Moore, A., Rhinehardt, K. L., Van Beveren, E., Yang, B., Tapia, A., Han, J., Harrison, S. H., & Graves, J. L. Jr. (2021). Too much of a good thing: Adaptation to iron(II) intoxication in *Escherichia coli. Evolution, Medicine, and Public Health, 9*(1), 53–67. https://doi.org/10.1093/emph/eoaa051

Tobler, R., Hermisson, J., & Schlötterer, C. (2015). Parallel trait adaptation across opposing thermal environments in experimental *Drosophila melanogaster* populations. *Evolution, 69*(7), 1745–1759. https://doi.org/10.1111/evo.12705

Todd, E. V., Black, M. A., & Gemmell, N. J. (2016). The power and promise of RNA-seq in ecology and evolution. *Molecular Ecology, 25*(6), 1224–1241. https://doi.org/10.1111/mec.13526

Tong, K., Datta, S., Cheng, V., Haas, D. J., Gourisetti, S., Yopp, H. L., Day, T. C., Lac, D. T., Khalil, A. S., Conlin, P. L., Bozdag, G. O., & Ratcliff, W. C. (2025). Genome duplication in a long-term multicellularity evolution experiment. *Nature, 639*(8055), 691–699. https://doi.org/10.1038/s41586-025-08689-6

Travisano, M., & Lenski, R. E. (1996). Long-term experimental evolution in *Escherichia coli*, IV. Targets of selection and the specificity of adaptation. *Genetics, 143*(1), 15–26. https://doi.org/10.1093/genetics/143.1.15

Travisano, M., Mongold, J. A., Bennett, A. F., & Lenski, R. E. (1995). Experimental tests of the roles of adaptation, chance, and history in evolution. *Science, 267*(5194), 87–90. https://doi.org/10.1126/science.7809610

Turner, C. B., Blount, Z. D., Mitchell, D. H., & Lenski, R. E. (2023). Evolution of a cross-feeding interaction following a key innovation in a long-term evolution experiment with *Escherichia coli. Microbiology (Reading), 169*(8), 001390. https://doi.org/10.1099/mic.0.001390

Turner, C. B., Wade, B. D., Meyer, J. R., Sommerfeld, B. A., & Lenski, R. E. (2017). Evolutionary changes in organismal stoichiometry during a long-term experiment with *Escherichia coli. Royal Society Open Science, 4*(7), 170497. https://doi.org/10.1098/rsos.170497

Turner, P. E., & Chao, L. (1998). Sex and the evolution of intrahost competition in RNA virus phi6. *Genetics, 150*(2), 523–532. https://doi.org/10.1093/genetics/150.2.523

Turner, P. E., & Chao, L. (1999). Prisoner's dilemma in an RNA virus. *Nature*, *398*(6727), 441–443. https://doi.org/10.1038/18913

Turner, T. L., Stewart, A. D., Fields, A. T., Rice, W. R., & Tarone, A. M. (2011). Population-based resequencing of experimentally evolved populations reveals the genetic basis of body size variation in *Drosophila melanogaster*. *PLoS Genetics*, *7*(3), e1001336. https://doi.org/10.1371/journal.pgen.1001336

Tyler, A. D., Mataseje, L., Urfano, C. J., Schmidt, L., Antonation, K. S., Mulvey, M. R., & Corbett, C. R. (2018). Evaluation of Oxford Nanopore's MinION sequencing device for microbial whole genome sequencing applications. *Scientific Reports*, *8*(1), 10931. https://doi.org/10.1038/s41598-018-29334-5

van Dijk, E. L., Jaszczyszyn, Y., Naquin, D., & Thermes, C. (2018). The third revolution in sequencing technology. *Trends in Genetics*, *34*(9), 666–681. https://doi.org/10.1016/j.tig.2018.05.008

Vasi, F., Travisano, M., & Lenski, R. E. (1994). Long-term experimental evolution in *Escherichia coli*. II. Changes in life-history traits during adaptation to a seasonal environment. *The American Naturalist*, *144*(3), 432–456. https://doi.org/10.1086/285685

Velicer, G. J., Kroos, L., & Lenski, R. E. (1998). Loss of social behaviors by *Myxococcus xanthus* during evolution in an unstructured habitat. *Proceedings of the National Academy of Sciences of the United States of America*, *95*(21), 12376–12380. https://doi.org/10.1073/pnas.95.21.12376

Velicer, G. J., Kroos, L., & Lenski, R. E. (2000). Developmental cheating in the social bacterium *Myxococcus xanthus*. *Nature*, *404*, 598–601. https://doi.org/10.1038/35007066

Velicer, G. J., & Lenski, R. E. (1999). Evolutionary trade-offs under conditions of resource abundance and scarcity: experiments with bacteria. *Ecology*, *80*(4), 1168–1179. https://doi.org/10.1890/0012-9658(1999)080[1168:ETOUCO]2.0.CO;2

Velicer, G. J., Raddatz, G., Keller, H., Deiss, S., Lanz, C., Dinkelacker, I., & Schuster, S. C. (2006). Comprehensive mutation identification in an evolved bacterial cooperator and its cheating ancestor. *Proceedings of the National Academy of Sciences of the United States of America*, *103*(21), 8107–8112. https://doi.org/10.1073/pnas.0510740103

Velicer, G. J., & Yu, Y. N. (2003). Evolution of novel cooperative swarming in the bacterium *Myxococcus xanthus*. *Nature*, *425*(6953), 75–78. https://doi.org/10.1038/nature01908

Vlachos, C., Burny, C., Pelizzola, M., Borges, R., Futschik, A., Kofler, R., & Schlötterer, C. (2019). Benchmarking software tools for detecting and quantifying selection in evolve and resequencing studies. *Genome Biology*, *20*, 169. https://doi.org/10.1186/s13059-019-1770-8

Walker, D. W., McColl, G., Jenkins, N. L., Harris, J., & Lithgow, G. J. (2000). Evolution of lifespan in *Caenorhabditis elegans*. *Nature*, *405*(6784), 296–297. https://doi.org/10.1038/35012693

Walsh, B., & Lynch, M. (2018). *Evolution and selection of quantitative traits*. Oxford University Press. Oxford, UK.

Waters, S. M., Zeigler, D. R., & Nicholson, W. L. (2015). Experimental evolution of enhanced growth by *Bacillus subtilis* at low atmospheric pressure: Genomic changes revealed by whole-genome sequencing. *Applied and Environmental Microbiology*, *81*(21), 7525–7532. https://doi.org/10.1128/AEM.01690-15

Wattiaux, J. M. (1968a). Cumulative parental age effects in *Drosophila subobscura*. *Evolution*, *22*(2), 406–415. https://doi.org/10.2307/2406539

Wattiaux, J. M. (1968b). Parental age effects in *Drosophila pseudoobscura*. *Experimental Gerontology*, *3*(1), 55–61. https://doi.org/10.1016/0531-5565(68)90056-9

Wick, R. R., Judd, L. M., & Holt, K. E. (2023). Assembling the perfect bacterial genome using Oxford Nanopore and Illumina sequencing. *PLOS Computational Biology*, *19*, e1010905. https://doi.org/10.1371/journal.pcbi.1010905

Wielgoss, S., Barrick, J. E., Tenaillon, O., Wiser, M. J., Dittmar, W. J., Cruveiller, S., Chane-Woon-Ming, B., Médigue, C., Lenski, R. E., & Schneider, D. (2013). Mutation rate dynamics in a bacterial population reflect tension between adaptation and genetic load. *Proceedings of the National Academy of Sciences of the United States of America*, *110*(1), 222–227. https://doi.org/10.1073/pnas.1219574110

Wiser, M. J., Ribeck, N., & Lenski, R. E. (2013). Long-term dynamics of adaptation in asexual populations. *Science*, *342*(6164), 1364–1367. https://doi.org/10.1126/science.1243357

Woods, R., Schneider, D., Winkworth, C. L., Riley, M. A., & Lenski, R. E. (2006). Tests of parallel molecular evolution in a long-term experiment with *Escherichia coli*. *Proceedings of the National Academy of Sciences of the United States of America*, *103*(24), 9107–9112. https://doi.org/10.1073/pnas.0602917103

Wright, S. (1931). Evolution in Mendelian populations. *Genetics*, *16*(2), 97–159. https://doi.org/10.1093/genetics/16.2.97

Wright, S. (1932). The roles of mutation, inbreeding, crossbreeding and selection in evolution. In *Proceedings of the Sixth International Congress on Genetics*, *1*, 356–366.

Wright, S. (1977). *Evolution and the genetics of natural populations. Vol. 3: Experimental results and evolutionary deductions*. University of Chicago Press. Chicago, IL, USA

Xu, T., Gong, Y., Su, X., Zhu, P., Dai, J., Xu, J., & Ma, B. (2020). Phenome-genome profiling of single bacterial cell by Raman-activated gravity-driven encapsulation and sequencing. *Small, 16*(30), e2001172. https://doi.org/10.1002/smll.202001172

Yang, J., Manolio, T. A., Pasquale, L. R., Boerwinkle, E., Caporaso, N., Cunningham, J. M., de Andrade, M., Feenstra, B., Feingold, E., Hayes, M. G., Hill, W. G., Landi, M. T., Alonso, A., Lettre, G., Lin, P., Ling, H., Lowe, W., Mathias, R. A., Melbye, M., . . . Visscher, P. M. (2011). Genome partitioning of genetic variation for complex traits using common SNPs. *Nature Genetics, 43*(6), 519–525. https://doi.org/10.1038/ng.823

Yengo, L., Vedantam, S., Marouli, E., Sidorenko, J., Bartell, E., Sakaue, S., Graff, M., Eliasen, A. U., Jiang, Y., Raghavan, S., Miao, J., Arias, J. D., Graham, S. E., Mukamel, R. E., Spracklen, C. N., Yin, X., Chen, S.-H., Ferreira, T., Highland, H. H., . . . Hirschhorn, J. N. (2022). A saturated map of common genetic variants associated with human height. *Nature, 610*(7933), 704–712. https://doi.org/10.1038/s41586-022-05275-y

You, S. T., Jhou, Y. T., Kao, C. F., & Leu, J. Y. (2019). Experimental evolution reveals a general role for the methyltransferase Hmt1 in noise buffering. *PLoS Biology, 17*(10), e3000433. https://doi.org/10.1371/journal.pbio.3000433

Yoxen, E. (1982). Giving life a new meaning: The rise of the molecular biology establishment. In N. Elias, H. Martins, & R. Whitley (Eds.), *Scientific establishments and hierarchies. Sociology of the sciences: A yearbook, Vol. 6* (pp. 123–143). Springer. Dordrecht, Netherlands. https://doi.org/10.1007/978-94-009-7729-7_5

Yu, Y. N., Kleiner, M., & Velicer, G. J. (2016). Spontaneous reversions of an evolutionary trait loss reveal regulators of a small RNA that controls multicellular development in Myxobacteria. *Journal of Bacteriology, 198*(21), 3142–3151. https://doi.org/10.1128/JB.00389-16

Zeyl, C., & Bell, G. (1997). The advantage of sex in evolving yeast populations. *Nature, 388*(6641), 465–468. https://doi.org/10.1038/41312

Zhou, W. M., Yan, Y. Y., Guo, Q. R., Ji, H., Wang, H., Xu, T. T., Makabel, B., Pilarsky, C., He, G., Yu, X. Y., & Zhang, J. Y. (2021). Microfluidics applications for high-throughput single cell sequencing. *Journal of Nanobiotechnology, 19*(1), Article 312. https://doi.org/10.1186/s12951-021-01045-6

Index